Hip And Knee Surgery:

A Patient's Guide To Hip Replacement, Hip Resurfacing, Knee Replacement, & Knee Arthroscopy

Robert Edward Kennon, M.D.

THIRD EDITION

ISBN 978-1-4357-0732-0

NOTICE

Every effort has been taken to confirm the accuracy of the information presented and to describe generally accepted practices. However, many practices differ from surgeon to surgeon, and those described here are common practice by the author and may differ from other surgeons. The author cannot be held responsible for errors or for any consequences arising from the use of the information contained herein, and makes no warranty, expressed or implied, with respect to the contents of the publications. In no circumstances should the content be used to substitute for the advice and instructions of a physician who is familiar with a patient's condition and responsible for providing treatment and education to the patient. It is the responsibility of the treating physician, relying on experience and knowledge of the patient, to determine the best treatment for the patient. Neither the publisher nor the author assumes any responsibility for any injury and or damage to persons or property.

TABLE OF CONTENTS

Introduction 1

PART I – THE HIP --

1 – Hip Anatomy and its Associated Problems 6

2 – Diseases of The Hip 10

3 – Diagnosing Hip Disease 17

4 – Nonoperative Treatment of Hip Arthritis 22

5 – Total Hip Replacement (Arthroplasty) 26

6 – Hip Resurfacing 31

7 – Surgical Approaches Used For Total Hip Replacement And Resurfacing 35

8 – Surgical Alternatives To Hip Replacement or Resurfacing 40

9 – Hip Replacement Prosthesis Designs 44

10 – Hospitalization For Hip Replacement or Resurfacing 51

11 – Home Exercises After Hip Replacement or Resurfacing 54

12 – Life After Hip Replacement or Resurfacing 64

PART II – THE KNEE ---

13 – Knee Anatomy and its Associated Problems 70

14 – Diseases of The Knee 75

15 – Diagnosing Knee Disease 83

16 – Nonoperative Treatment of Knee Arthritis 87

17 – Knee Arthroscopy 92

18 – Total and Partial Knee Replacement (Arthroplasty) 97

19 – Surgical Approaches Used For Knee Replacement 103

20 – Other Surgical Alternatives To Knee Replacement 107

21 – Knee Replacement Prosthesis Designs 110

22 – Hospitalization For Knee Replacement 115

23 – Home Exercises After Knee Replacement 118

24 – Life After Knee Replacement 123

PART III – BEFORE & AFTER JOINT REPLACEMENT SURGERY ----------------------------------

25 – Office Visit Before Surgery and Initial Evaluation 130

26 – Deciding On Surgery 133

27 – Scheduling Surgery and What to Expect With Preoperative Testing 136

28 – Blood Donation Before Total Joint Surgery 140

29 – The Team Taking Care of You 143

30 – Hospital Admission And Your Medications Before Surgery 147

31 – What to Expect The Morning Of Surgery 150

32 – Anesthesia 153

33 – What to Expect After Joint Replacement – Getting Around, Physical Therapy, Medications 158

34 – Preventing Thromboembolism After Joint Replacement And Use Of Anticoagulation 162

35 – What to Expect After Leaving the Hospital – The First 3 Weeks   168

36 – Complications of Surgery   172

PART IV – REVISION JOINT REPLACEMENT SURGERY --

37 – Revision Hip And Knee Surgery   190

APPENDICES --

APPENDIX I – RESOURCES   196

APPENDIX II – DENTAL CARE RECOMMENDATIONS   197

APPENDIX III – CHECKLISTS   199

APPENDIX IV – ABOUT THE AUTHOR AND THE PRACTICE   201

APPENDIX V – CONTACT US   202

GLOSSARY   203

INDEX   212

Acknowledgment

I wish to thank Dr. Kristaps Keggi and Dr. John Keggi, both of whom have been mentors, colleagues, and friends, and who I still learn new things from with each passing year.

Introduction

This book is intended for our patients. It is dedicated to them and to helping them understand their joint disease, evaluate their options, make better decisions regarding their care, and ultimately, to get better.

As a surgeon, I have found that probably only 25% of my time is spent actually helping patients in the operating room. The majority of my time is spent teaching patients about their health problems and options, both in the hospital and in the office, sketching diagrams at the bedside or in the exam room, and conveying information about their condition and options. I enjoy both parts of patient care greatly, but I often find myself conveying the same information repeatedly. It occurred to me that it would be tremendously helpful to try to anticipate and answer patients' questions in a written format or book, particularly after a number of patients have repeatedly requested or suggested just such an information source.

The other source of motivation for me to write this book is the knowledge that the best patient outcomes occur when patients are both well-motivated to get better and have a good understanding of what to expect.

The primary goal of this book is to put complex information in the most accessible and portable form possible, so that patients and their families may refer to it at once whenever they have a question, rather than scratch their head and try to remember to ask the surgeon during relatively quick encounters in the office or hospital. Additionally, the natural anxiety that patients feel when contemplating surgery may be decreased by reading through a clear and simple explanation of the steps involved with preoperative consultation, hospital admission, surgery, and rehabilitation.

There are a few very important things to remember before reading this book, however.

The guidelines that are put forth here are the typical recommendations and suggestions (such as postoperative exercises) that my associates and I use in our own practice. Not all surgeons will use the same surgical approaches, techniques, rehabilitation recommendations, and implants, and undoubtedly patients should refer to the advice of their surgeon first. If you are having surgery done by another surgeon, you may be advised of different activity limitations, medication regimens, etc., and certainly a book cannot replace the dialog that occurs between doctor and patient. You will need to obtain your post-operative instructions and medical advice from your orthopaedic surgeon, although you may be better equipped to ask questions and understand the reasons behind what you are asked to do and not do if you research joint replacement surgery on your own.

For my own patients, most of this book will have a familiar ring to it. It is helpful to have the usual guidelines and information here in one place.

This book is arranged into several large sections: the first section focuses on hip diseases and surgery, the second section focuses on knee diseases and surgery, and the third discusses hospitalization and medical issues that are common to both types of joint replacement. The fourth section of the book is an overview of revision joint replacement surgery.

This entire book is centered around the subspecialty of orthopaedics known as ***adult reconstruction***. This refers to partial and total joint replacements, resurfacing procedures, and related procedures such as osteotomies, arthroscopy, and also the nonoperative management of these problems. There is some overlap with other subspecialty areas of orthopaedics, such as sports medicine or trauma, and where applicable procedures such as knee arthroscopy or hip fractures are discussed.

There is an index at the end of the book to facilitate quick reference to specific areas, and a short orthopaedic glossary is also included so that the reader can quickly look up an unfamiliar term that he or she might have heard. Several other useful appendices are included for quick reference.

Some excellent medical illustrations reproduced in this work, where noted, have been used with the gracious permission of Smith & Nephew. The rest are my own work, and Simon the computer model has posed tirelessly for the illustrations in this book. Simon the computer model is sometimes nude but, as you will notice, he is missing a few anatomical features so as not to offend any readers.

Every effort has been made to keep the text and explanations in clear language that is intended for the reader who has curiosity and interest in the subject. Although the unique mix of physiology and engineering in orthopaedic surgery can sometimes seem overly complex, most things can be readily explained in fairly straightforward terms by focusing on the important concepts rather than details. If something is unclear, then make a note and ask your surgeon about it.

The first section begins with a quick overview of the anatomy of the hip joint. This book is designed so that it need not be read cover to cover, and you should feel free to skip to any section of interest.

I hope that this book helps you. If this book assists in making the process of facing orthopaedic surgery easier, helps my patients understand their options better, and increases my patients' knowledge of the subject, then it has accomplished what it was meant to.

Warm regards,

Robert Edward "Ted" Kennon, M.D.

Simon the computer model posed tirelessly for the illustrations used throughout this book. Where applicable, the observant reader will notice that poor Simon is missing a few anatomical features so as not to offend any readers.

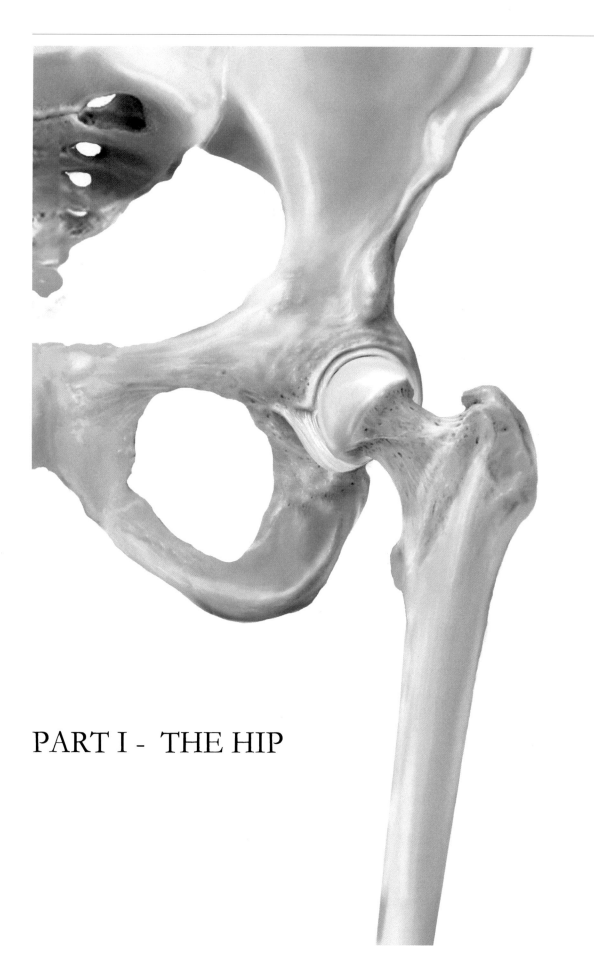

PART I - THE HIP

Chapter 1 - Hip Anatomy and its Associated Problems

The hip is a ball and socket joint. At its simplest, the hip functions as a ball bearing. This design allows the hip to flex, extend, move from side to side (called abduction and adduction), and rotate (called internal rotation and external rotation). Although not quite as complex as the knee, the hip joint does its job wonderfully and allows us to get around with surprising agility (when it works properly!).

Bones And Cartilage

Normally the "ball" (femoral head) and "socket" (acetabulum) are very smooth, and both sides of the joint (as with most joints in the body) are covered with a smooth layer of cartilage that is called articular cartilage. This layer is about $1/8^{th}$ of an inch thick on both sides of the joint. It is analogous to having Teflon that coats a frying pan; similarly, if that smooth coating is scraped or worn off, it leaves the underlying bone surface exposed. This is where the term "bone on bone" arthritis comes from.

There is a cartilage "bumper" or gasket that surrounds the rim of the socket, called the labrum. This gasket can develop tears and cause problems, typically in the form mechanical difficulties and pain from the torn piece (a labral tear).

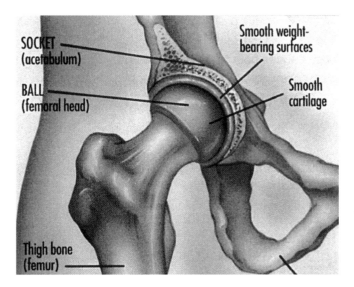

Figure 1-1. Normal hip anatomy and cartilage. *(courtesy of Smith & Nephew, reprinted with permission)*

The socket itself (acetabulum) can sometimes be formed incorrectly, leading to congenital problems that may be apparent as an infant or that go undetected until adulthood. Most commonly, this happens if the socket is too shallow, and the hip has a tendency to "pop out" or dislocate as an infant. Even if the hip does not ever completely dislocate, the abnormal shape of a shallow acetabulum (called hip dysplasia) can cause problems years or even decades later. This is a common reason for adults to wear out their hip joint at an early age and require surgery.

Arthritis simply means there is an inflammation of the joint (there are multiple types of arthritis, as will be discussed in the next chapter), and this accompanies the loss of the smooth cartilage surface.

As the surface becomes increasingly rough with loss of the smooth cartilage that coats the joint, other changes begin to occur as well. The body may react by forming large spurs around the joint, called osteophytes. The underlying bone surface becomes more dense and hard in order to resist the forces against the exposed bone surface. These dense changes are called subchondral sclerosis. The bone around the joint may also commonly develop cysts that fill with joint fluid, known as subchondral cysts. Eventually, the bone surfaces begin to erode as arthritis worsens.

Blood Supply

The "ball" of the femoral head itself is supplied with a blood supply from several sources. Some of these include the circumflex arteries around the base of the femoral neck. Injuries to this blood supply such as a traumatic hip dislocation, or clotting disorders that prevent blood flow in this region such as Sickle Cell Disease, can lead to a loss of the bone called avascular necrosis (discussed in the next chapter).

The hip joint is surrounded by a tough covering called the capsule. Injections into the hip joint are actually injections inside this capsule. Similarly, infections often involve the joint fluid and space within the capsule. If this capsule becomes contracted and tight, either with disuse or aging, it can lead to a flexion contracture of the hip where it is difficult to fully lay the leg flat while laying down. This often requires a release at the time of hip surgery to restore range of motion. Loose bodies, or small nuggets of bone or cartilage (sometimes called "joint mice") are usually found in this joint space within the capsule.

Bursa

The prominent bony area over the side of the hips is called the greater trochanter (and there is in fact a lesser trochanter also, but it is located deep on the inside of the thigh). This area can be involved in hip fractures, particularly when someone lands directly on to their side. More commonly, many patients develop pain in this area due to bursitis. The bursa is a small sac that fits between muscle layers, similar to two layers of plastic wrap with olive oil in between. These sacs are located all over the body, usually over joints where they assist in allowing muscle layers to slide smoothly over one another. If the sac becomes inflamed, it is referred to as bursitis, and this is commonly seen over the side of the hip, shoulders, elbows, and the front of the knees.

The most common cause of pain over the side of the hip is from trochanteric bursitis, which causes pain when pressing on or laying on the side of the hip.

Muscles

A number of muscles attach to the femur and the pelvis in this region, and the anatomy of all of these becomes more complex than we will be discussing here. However, a few important ones bear mentioning. The psoas muscle originates all the way up near the spine before passing down through the pelvis and attaching to a bony prominence on the inside of the thigh, the lesser trochanter. The psoas is primarily responsible for flexing the hip upward. It can sometimes become inflamed or cause pain (tendinitis). A group of large muscles over the side and back of the hip comprise the gluteus muscles, which are involved in pulling the hip backwards (extension) and out to the side (abduction). In the front of the hip, the tensor fascia lata, sartorius, and rectus femoris are involved in pulling the leg up into flexion and pass over the front, or anterior, of the thigh. A group of small muscles attach deep and behind the trochanter, collectively called the short external rotators (such as the piriformis, quadratus, obturator externus), responsible for turning the hip outward. Sometimes the piriformis can compress the sciatic nerve and cause pain (piriformis syndrome). All of these muscle groups will be revisited when discussing different surgical approaches in later chapters.

Nerves and Blood Vessels

For those readers that are interested, there are a number of nerves and blood vessels around the hip. While these are important to your surgeon, a detailed knowledge of the neurovascular anatomy is not required to understand most surgeries and diseases unless we are discussing specific injuries or problems with these structures, and so just a few of the larger and more important structures are mentioned here. More detailed discussions regarding nerve injuries are included in later chapters.

In the front of the hip, the femoral artery, nerve, and vein travel together in a neurovascular bundle along the groin and inner thigh. A branch of the femoral nerve, called the lateral femoral cutaneous nerve, can be involved in numbness or pain over the front of the thigh (sometimes called *meralgia parasthetica* when it is severe). It is common to have some numbness or tingling in that region for months after surgery, but it is not a motor nerve and does not affect muscle strength.

The sciatic nerve is a very large nerve that runs down the back of the leg. It can be injured particularly in surgical approaches that enter from the back of the hip, and it contains both motor and sensory fibers. In the worst case, an injury to the sciatic nerve can result in a sciatic palsy, or "foot drop," in which it is difficult or impossible to raise the foot upwards from the ankle. Sometimes the sciatic nerve can be affected near the spine where it exits the spinal cord, typically from a bulging disc (herniated or "slipped" disc), and this can cause shooting pains and numbness down the leg. These spine problems at the level of the sciatic nerve and nearby spinal levels can sometimes cause leg or hip pain that can mimic hip problems, necessitating special tests to differentiate and determine where the pain is actually coming from. Similarly, another nerve runs along the inner aspect of the hip joint and down to the knee – the obturator nerve – which often accounts for why patients with hip problems sometimes present with initial complaints of knee pain rather than hip pain.

The large veins around the hip are usually far from the area of surgery, but these are sometimes significant if a blood clot forms within them after surgery or injury, known as a deep venous thrombosis (DVT). It often forms farther down the leg, but the backup in blood flow leads to significant swelling and pain in the leg, especially the calf. The clot itself is not usually dangerous, but a small percentage of the time, a piece of the clot can break off and travel with the circulation to the heart and lungs, which can be serious. A blood clot that travels through the right side of the heart and then lodges in the lungs is called a pulmonary embolus (PE), which can be fatal. This is why most surgeons prescribe some sort of "blood thinner" or anticoagulant after surgery, such as aspirin, heparin (or a related low-molecular weight heparin, such as enoxaprin injections), or warfarin. This is an important topic, and an entire chapter is devoted to it.

A Tough Ball Bearing

The hip joint carries large loads, typically of more than a million steps per year for most active adults. That is a lot of potential wear and tear, but surprisingly, it actually works very well until some underlying process disrupts or overcomes its ability to make routine repairs. Like a mechanical bearing in a piece of machinery, the wear and tear usually takes a long time to gradually become apparent, and this process is called arthritis.

Key Points For This Chapter:

- The hip is a ball and socket joint formed by the femur (ball) and the acetabulum (socket)

- The bone surfaces are covered with about 1/8th inch of articular cartilage; this is worn away with arthritic changes and leads to the common description of "bone on bone" degenerative changes.

- The cartilage bumper around the socket is called the labrum, which can develop tears

- Not all hip pain/problems come from the hip joint itself (e.g., bursitis or sciatica).

Chapter 2 - Diseases of The Hip

The first chapter mentioned some of the potential problems that can be associated with specific anatomical features. Some categories of hip problems are broader, and these need specific discussion. The different types of arthritis are broad categories of disease, and a full discussion of all of the types of arthritis that have been described could fill several textbooks alone. In fact, a rheumatologist is a specialized internist who treats these problems primarily with medicine (nonsurgical treatment) and accounts for an entire medical subspecialty. However, we will focus on some of the more common and important types which together account for the vast majority of orthopaedic surgery patients.

Hip Pain

This chapter discusses the most common types of arthritis and other common causes of hip pain, such as trochanteric bursitis or changes after fractures.

True hip joint pain usually presents as groin pain, although occasionally some patients will have primarily buttock or knee pain. Movement of the hip joint typically becomes limited and activities such as putting on socks or clipping toenails becomes increasingly difficult. A limp begins to develop. A small percentage of patients with serious hip disease will actually present with knee pain, which may be due to referred pain because of the overlapping nerve supply. Some types of pain felt around the hip joint may originate from the muscles or other soft tissues outside of the joint itself, such as bursitis. It is also possible that pain in the hip may be completely unrelated to the hip, caused by a lumbar radiculopathy or sciatica (a pinched nerve originating from the spine, usually because of a bulging herniated disc), gynecologic sources, or even a hernia.

Osteoarthritis

This is the "wear and tear" form of arthritis that most patients have. It is often explained to patients as being a gradual wear of the joint ("it's not the years, it's the mileage"), although it typically does not become clinically significant until middle age or later. Nearly all adults over forty will demonstrate at least some osteoarthritis in their joints even though it may not cause pain or problems for many years.

The cartilage coating over the joints wears away, eventually exposing the underlying bony surfaces (analogous to scraping away the Teflon in the frying pan). As this occurs, the body reacts by forming large bone spurs (called osteophytes), extra joint fluid (which may cause an effusion, or joint fluid accumulation), hard underlying bone surfaces (subchondral sclerosis), or cysts around the joint. Eventually, the hip joint begins to resemble a cauliflower more than a smooth, round ball. This usually causes pain in the groin, although it may radiate to the knee, buttock, or side of the hip. The joint becomes progressively stiff, so that it becomes difficult to put on shoes and socks, clip toenails, get up out of a chair, etc.

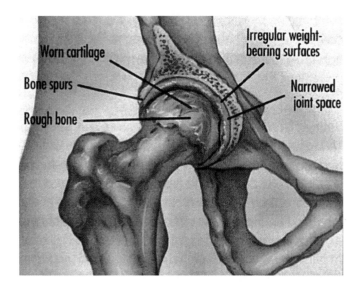

Figure 2-1. Degenerative changes from osteoarthritis. *(courtesy of Smith & Nephew, reprinted with permission)*

There is convincing evidence that many patients with osteoarthritis are prone to get it based on their genetics. Some studies involving identical twins suggest that occupation and other factors may not play as great a role as previously thought, and the tendency to develop severe osteoarthritis runs in families. A study recently showed that in identical twins, both usually had similar patterns and severity of osteoarthritis later in life, even if one became a heavy laborer and the other worked at a desk job.

Related forms of osteoarthritis include post-traumatic arthritis (arthritis that forms after an old injury, usually a fracture or a dislocation of the hip), late sequelae or consequences of prior diseases (patients who had slipped capital epiphyses or Legg-Calve-Perthes disease as children), and congenital defects such as a shallow hip socket (hip dysplasia).

Rheumatoid Arthritis

Although many patients may describe their joint pain as "rheumatism," rheumatoid arthritis is a special form of arthritis in which the body's own immune system attacks the joints. This leads to large, swollen joints that are painful and frequently reddened and warm. Rather than attacking a few joints, most rheumatoid arthritis patients have pain in many joints.

Blood tests often can detect or confirm the presence of rheumatoid arthritis (although they are not completely accurate), and it is usually diagnosed by the family physician or a rheumatologist when a patient's symptoms are suspicious. Untreated, the inflammation within the joints leads to destruction of the cartilage in the joint, and the inflammation caused by the body's immune system attacking its own tissues can also have consequences such as tendon ruptures and hand deformities. Newer rheumatologic medications to suppress the autoimmune disease have led to a remarkable decrease in the number and severity of rheumatoid patients that require orthopaedic surgical intervention. However, it remains a common source of joint problems and still accounts for many joint replacement surgeries each year.

Related forms of autoimmune arthritis also include Crohn's disease, ulcerative colitis, and psoriatic arthritis. Many patients are surprised to learn that psoriasis can lead to severe joint arthritis and destruction. Psoriasis is itself an autoimmune problem in which the body attacks itself, leading to eruptions in the skin that are typically associated with the disease. Like rheumatoid arthritis, it is often managed with medications that prevent the autoimmune response.

Avascular Necrosis (Osteonecrosis)

A potentially devastating condition which can affect patients of any age is avascular necrosis, or AVN, of the hip joint, which usually begins as the gradual onset of groin pain that worsens over time. In recent years, this has also been termed osteonecrosis. Essentially it is a syndrome in which the bone in the femoral head (the ball at the top of the thigh bone) begins to die. Usually, it is because that area of bone does not receive enough blood supply, due to a number of potential causes.

Avascular necrosis can be the result of trauma, clotting diseases, Sickle Cell disease, steroid use (usually long term use of oral prednisone), chemotherapy or radiation treatment for cancer, environmental factors (deep sea diving or working in a deep mine, where barometric changes chronically can lead to this disorder), metabolic diseases (such as Gaucher's Disease), and most commonly due to chronic, heavy alcohol use.

However, despite all of the numerous factors identified with causing avascular necrosis, nearly half of all patients have no identifiable risk factors. Many surgeons agree there is about a 50% lifetime risk of the opposite hip eventually becoming affected after developing the disease in one hip. It is one of the most common reasons for hip replacement surgery in young patients, either due to intractable severe pain or from arthritis that results after the bone in the femoral head has died and the round portion of the ball collapses.

Avascular necrosis can also affect other joints in the body, notably the knee (in the distal end of the femur), rarely the shoulder, and the small bones of the wrist (Kienbock's disease and Preiser's disease) and foot.

Septic Arthritis

Septic arthritis occurs when a joint becomes infected. This can happen most often with infants (because of the way the developing blood supply feeds the joints, making it easier for a systemic infection to seed a joint space), people who use intravenous drugs (such as heroin), and with patients who have immune system impairments (such as HIV or AIDS). Most surgeons will not consider joint replacement surgeries for patients with a history of IV drug abuse because of the potential of infecting an artificial joint if the patient engages in IV drug use.

Often, the source of the infection is obvious, such as a chronic diabetic ulcer or infection along the leg that has not healed, a severe urinary tract infection, recent unrelated surgery, or recent systemic illness. However, sometimes the source cannot be identified, and it is assumed to have occurred through transient bacteremia (temporary presence of bacteria in the bloodstream, as occurs every day shortly after brushing your teeth).

Patients with septic arthritis of the hip from any source are usually quite ill, with high fevers and severe groin pain with any movement of the hip. The infection can very quickly destroy the cartilage if left untreated (or destroy the bone fixation to an implanted artificial joint). An infection with pus inside any joint is a surgical emergency and usually requires immediate surgery to wash out the joint.

Trauma and Fractures

There are a number of different fractures that can involve the hip and pelvis, and many of these require surgery to fix and stabilize the bone. Hip fractures are common in elderly patients, who often have osteoporosis (or weak bones) and sustain fractures with falls. There are several different types of hip fractures.

Femoral neck fractures (also called subcapital fractures) occur when the femur breaks just below the ball of the femur. This can occur as the result of a fall or sometimes can gradually occur over time as the result of a stress fracture. If the fracture does not displace, this can sometimes be treated with percutaneous pinning. Pinning is very quick and is a minimal surgery compared to other repairs, but the patient must be able to follow directions and limit weightbearing until the fracture has healed, which can be a problem for nursing home patients without the strength or the mental presence to stay off of the hip. Displaced femoral neck fractures in which the ball falls off the neck of the femur are usually treated with partial or total hip replacements, depending on the activity level of the patient before the injury.

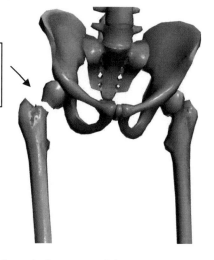

Femoral neck fracture (displaced)

Figure 2-2. A femoral neck fracture of the hip.

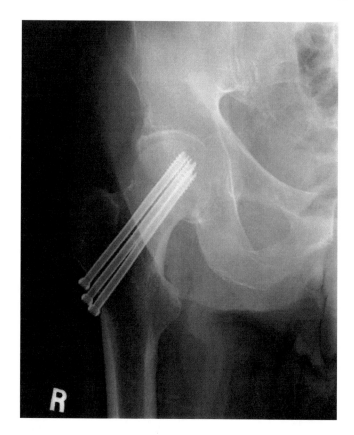

R

Figure 2-3. A femoral neck fracture that has been treated with percutaneous pinning. The fracture has healed 6 months after the injury.

Hip fractures that involve the portion below the ball and neck area, in the region of the prominent bony part of the thigh called the greater trochanter, are referred to as intertrochanteric fractures. These often occur from landing directly on the side of the hip. These are usually repaired with either a plate along the side of the femur and a screw going into the head of the femur or with a rod through the femur that is placed in a newer, less invasive surgery (known as an intramedullary hip screw). These methods are also used to repair subtrochanteric fractures, which occur closer to the region of the femoral shaft. Hip replacement is possible for these types of hip fractures but is usually not the first choice of treatment, as fractures in this region of the hip involve the area where the stem of the hip prosthesis is anchored. If hip replacement is used to treat these types of fractures, usually a special type of hip replacement known as a calcar replacing prosthesis is used. The most common reason to use a hip replacement for intertrochanteric or subtrochanteric hip fractures is after a previous fixation with a hip screw fails (usually due to screw cut-out through the soft, osteoporotic bone or because the patient did not follow weightbearing instructions/limitations).

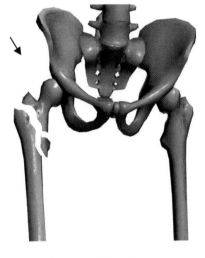

3 Part intertrochanteric fracture (displaced)

Figure 2-4. A 3 part intertrochanteric hip fracture.

Some fractures involve the socket (acetabulum) of the hip rather than the femur. Acetabular fractures vary in severity. Some fractures involving this socket are minor and are treated without surgery, managed with just limited weightbearing until the fracture heals. Some are more severe, particularly fractures that blow out enough of the socket that the ball cannot stay located and falls out of the joint (dislocates). These require that the socket either be repaired with screws and/or plates, or alternatively, the entire hip may be replaced.

Even when the above types of fractures have healed, most patients develop arthritis years later (known as post-traumatic arthritis). It is similar to osteoarthritis. Patients who previously had a hip dislocation from an accident also often develop arthritis

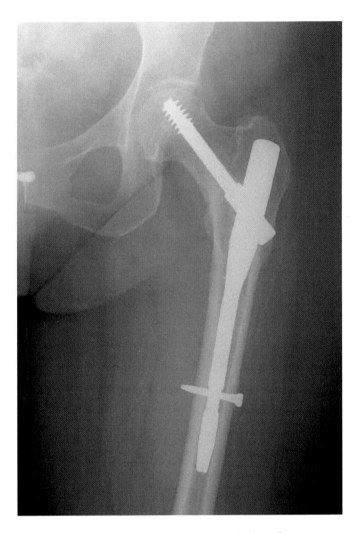

Figure 2-5. A healed intertrochanteric hip fracture treated with an intramedullary hip screw 6 months after the injury.

in the hip joint years afterwards. It is not uncommon for patients who have successfully healed from hip or acetabular (socket) fractures to eventually require hip replacement.

Tumors

Bone tumors that arise within the bone itself are rare, although many types have been described. It is far more common however to have metastatic cancer which spreads to the bone from other locations. Metastases may involve the femoral neck (the portion of the femur just below the ball), which can lead to fracture. Pain usually precedes a fracture, however, and bone scans and other tests usually reveal the tumor before fracture occurs. Sometimes a metal rod or pins are used to strengthen the involved area, and at other times a hip replacement may be used to make the patient more comfortable and preserve their ability to bear weight and walk. The five tumor types that most commonly are responsible for spreading to bone are breast cancer, lung cancer, thyroid cancer, prostate cancer, and kidney cancer.

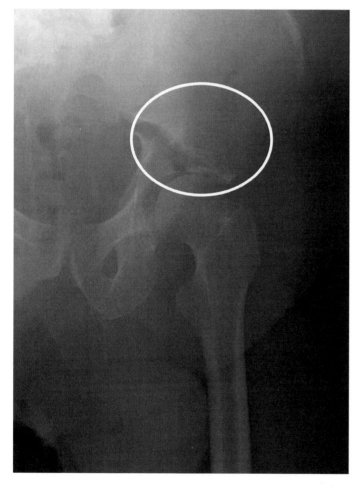

Figure 2-6. An acetabular (socket) fracture [circled area]. This injury occurred as a result of a motor vehicle accident.

Labral Tears

The hip socket (acetabulum) has a cartilage "bumper" or gasket that surrounds the rim. If this gasket tears, the torn cartilage may not heal on its own, leading to groin pain and mechanical symptoms such as locking and catching in the hip joint. Although many of these tears are managed conservatively and eventually heal on their own, some are debilitating and require surgery to remove the torn fragment of cartilage that is being caught inside the joint.

Trochanteric bursitis

The bursa (or sac) between muscle layers over the lateral side of the hip can become inflamed and painful. The hip pain is reproduced by pressing on the side of the hip where the bone is most prominent, and most patients report that it is uncomfortable to lay on the affected side (especially at night).

It is often associated with weight gain, injury, or repetitive inflammation through activities such as running. In a few cases, the pain can be severe enough that it causes a mild limp. This type of bursitis is very common, and many patients are relieved to learn that they do not require surgery. Instead, trochanteric bursitis is usually treated with weight loss, anti-inflammatory medications, muscle massage, stretching exercises / physical therapy, and an occasional steroid injection.

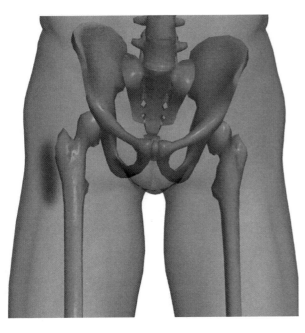

Figure 2-7. Trochanteric bursitis is an inflammation of the soft tissues over the side of the hip.

Ischial bursitis

Another related problem is ischial bursitis, which is less common than trochanteric bursitis but similarly affects a small bursa that is located posteriorly in the buttocks. Specifically, this is the bony prominence that you sit on, and it often results in point tenderness directly over that spot. This gives pain in the buttock area. It is usually treated in a similar manner to other types of bursitis, with anti-inflammatory medications and occasional steroid injections if severe.

Key Points For This Chapter:

- Osteoarthritis is the wear & tear type of arthritis

- Rheumatoid arthritis is caused by the body's immune system attacking the joints

- Other forms of arthritis include post-traumatic, infectious (septic), and late changes years after other diseases (Legg-Calve-Perthes Disease, Slipped Capital Femoral Epiphysis, Sickle Cell Anemia, etc.)

- Congenital hip dysplasia occurs when the socket did not form properly (usually too shallow) and can cause problems early on or later in life, depending on the severity.

- Avascular necrosis (osteonecrosis) is a common reason for hip destruction across all ages and has multiple causes, resulting in loss of the bone in the ball (femoral head)

- Hip fractures come in several different types and are treated with different types of surgeries (pinnings, plate & screws, intramedullary hip screws, partial or total hip replacements)

- Tumors in bone are rare. Metastatic cancer from other areas of the body is more common (e.g., breast, lung, prostate, thyroid, kidney) and may weaken the bone, requiring surgery

- Trochanteric bursitis is a very common cause of hip pain over the side of the hip, does not require surgery, and can usually be treated conservatively with physical therapy, anti-inflammatory medications, and steroid injections

- Ischial bursitis is a potential cause of posterior hip pain in the buttocks area

Chapter 3 - Diagnosing Hip Disease

Most hip diseases are diagnosed with a thorough history, a straightforward physical examination, and routine x-rays (radiographs). Blood tests and additional imaging tests, such as MRI, are not typically required for most diagnoses.

An orthopaedic surgeon will typically ask questions about the involved joint, activity levels, and symptoms. In most offices, an initial intake questionnaire usually covers most of the basic questions and medical history. The surgeon usually looks this over first, and x-rays may be ordered before or after examining the patient. Nearly all patients having hip or knee surgery will have an x-ray beforehand.

The history is often the most informative part of the interview. Surgeons usually ask about the location, severity, and frequency of the pain, along with what sorts of things bring it on and what makes it better. Specific questions with hip problems may deal with previous history, particularly if there has been any prior surgery or accidents, or with risk factors associated with certain disease processes (such as alcohol consumption when considering avascular necrosis).

The physical examination usually focuses on the affected joints themselves and adjacent joints, checking range of motion and function. Neurologic and vascular function are usually noted. There are many provocative tests and maneuvers used during a physical examination to further narrow down the particular source of the problem, such as a straight leg raise test for checking sciatic nerve problems.

Do not be surprised if your surgeon watches how you walk in and out of the office. Gait abnormalities are often very suggestive of the problem.

Radiographs (X-rays)

Plain x-rays of the hips are usually taken to evaluate for arthritis, fractures, congenital anomalies (such as hip dysplasia, or shallow sockets), tumors or metastatic disease, and other conditions. There are many things that the surgeon will be evaluating, often focusing on the appearance of the joint itself.

The cartilage that coats the surfaces of the joints is transparent on the x-ray, but if the gap is not apparent, "bone-on-bone" arthritis can be seen. Other features of arthritic joints include subchondral sclerosis (hardening of the underlying bone), osteophytes (spurs), loose bodies, and cysts in the bone. Fractures, tumors, and disease processes such as avascular necrosis are usually also readily visible on routine radiographs. These are also used to evaluate previous joint replacements.

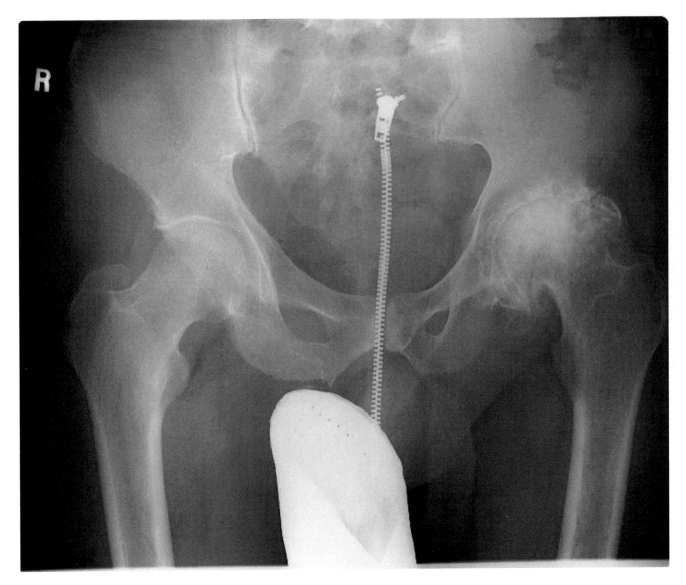

Figure 3-1. An x-ray showing severe degenerative changes in the patient's left hip (on your right, looking at the page). The opposite hip, in contrast, appears relatively normal with a rounded femoral head and a visible joint space made up of articular cartilage. *(The line in the middle of the x-ray is the zipper on his pants.)*

Magnetic Resonance Imaging (MRI)

MRI may sometimes be ordered to evaluate for soft tissue problems (such as muscle injury, evaluation of a soft tissue mass, etc.) or for bone marrow problems. It often will "light up" for increased water content, signaling edema and injury. Bone bruises and stress fractures show up in this manner. Avascular necrosis (osteonecrosis) is often diagnosed on MRI much earlier than when it appears evident on regular x-rays.

A torn rim of cartilage around the hip socket (a labral tear) is often visible on MRI, although sometimes an MRI arthrogram is required to see it well. This involves injecting the hip joint with a contrast dye in order to see tears more clearly. Intravenous contrast can be used also and is helpful for diagnosing tumors and infections with MRI.

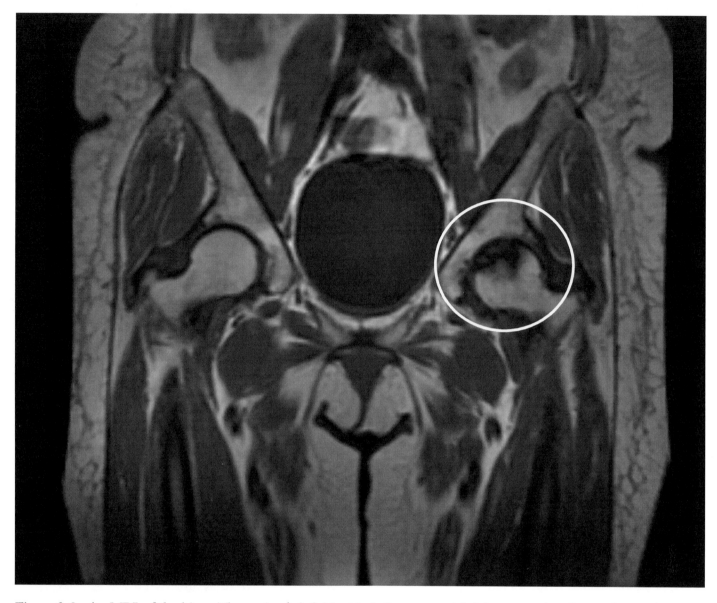

Figure 3-2. An MRI of the hips. The patient's left hip (circled, on your right) has extensive disease [avascular necrosis] when compared to the opposite hip. This would not be as easily detectable yet on routine x-rays.

Computed Tomography (CT)

A computed tomography (CT) scan uses many x-ray "slices" to examine cross sections of a body or limb. The patient lays on a table while moving through a ring that contains a spinning x-ray camera. While it does have applications in spine surgery and trauma (especially when examining complex pelvic fractures), CT scans are not typically used to evaluate for arthritis.

Nuclear Bone (Technecium) Scans

Bone scans involve administering a very small amount of radioactive material via an IV, then using a camera to view how it is taken up and moved by the tissues. Areas with high uptake, such as a tumor, infection, or fracture, will often "light up." This test is also useful for determining if an old hip or knee replacement is loosened from the bone, although it will

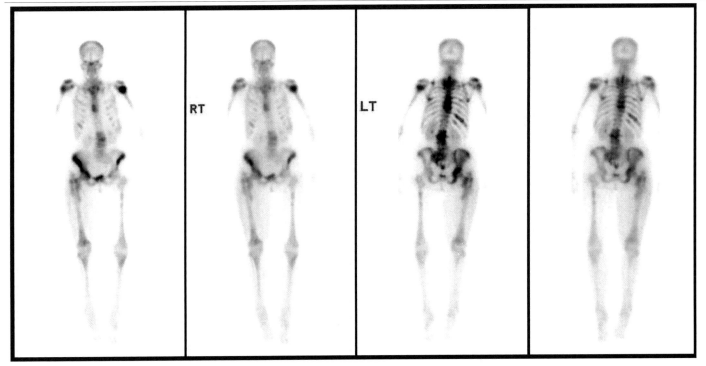

Figure 3-3. A Technicium bone scan. Areas of increased uptake (tumor) appear more dense in these images. This patient has metastatic breast cancer, involving the spine, left upper humerus, and right hip. The hip was about to fracture, and she underwent a hip replacement. Currently, she is alive and well, walking about on the hip replacement, with stable control of her cancer for years now with current oncology options.

provide a false positive if a bone scan is obtained within about 1 year or less of the surgery (because of normal bone remodeling).

White Blood Cell (WBC) Scans

A tagged white blood cell scan is a similar test to a bone scan, except white blood cells are taken from the body and "tagged" with a tiny amount of radioactive material. It is then re-injected, and the scanner shows where all of those tagged white blood cells congregate in order to localize an infection. This is used when trying to find an infection in the bone or around an artificial joint. It is usually used in conjunction after a routine technicium bone scan if infection is suspected. A related scan is a marrow (sulfur colloid) scan, which is also used to help diagnose and localize infections.

Ultrasound

An ultrasound examination uses sonic waves to make a picture of soft tissues. While it may be used to examine cysts or other soft tissue abnormalities (such as checking to see if an Achilles tendon is ruptured), its most common application in orthopaedics is to detect the presence of blood clots in deep veins.

Hip Aspiration (Arthrocentesis)

Sometimes it is useful to draw joint fluid out of the joint with a needle for laboratory analysis. This is most common in situations where there has been or there is suspicion for an infection in the joint itself. Usually the skin is anesthetized with a local anesthetic and the needle is placed into the joint under live x-ray. This is typically a quick procedure that is not much more uncomfortable than starting an I.V. The aspirated fluid is then sent to the laboratory for analysis to see if there is any evidence of bacteria or infection.

Hip Injection

Sometimes it is useful to diagnostically inject the hip joint with anesthetic and/or steroid in the same manner as for an aspiration (in fact, aspiration and injection are often performed together), but for very different reasons. This is most commonly done if there are multiple potential sources for the hip pain and/or it is difficult to sort out how much of the problem is originating from the hip itself.

A common scenario is the patient who has both an arthritic hip and sciatic pain originating from the spine. If the patient's pain improves significantly for a while, then this gives some idea of where the pain is coming from (e.g., it confirms that the pain is coming from within the hip joint). If not, then it is helpful to know and efforts can be focused elsewhere and on other causes of referred pain.

Hip injections are also performed for temporary pain relief, such as for a patient who needs a hip replacement but cannot (for medical or social reasons) proceed with the surgery for some time. The steroid injection often provides at least some pain relief that lasts from weeks to months.

Key Points For This Chapter:

- Most hip problems are diagnosed through the history and physical examination

- Radiographs (x-rays) are usually all that is needed to image the joint for most conditions

- MRI is used for examining soft tissues (like labral tears) or bone marrow (for detecting avascular necrosis or occult fractures)

- CT Scans are used to examine bone in detail, especially when 3D images are needed as in trauma cases

- Bone scans are used to detect loosening of old implants, stress fractures, or infections in the bone

- Tagged white cell (WBC) scans are used to detect and localize infection

- Ultrasound is used to examine tendons or to rule out blood clots in deep veins

- A hip aspiration is sometimes used to draw fluid out of the hip joint for testing, such as for infection.

- A hip injection may also be used to see if the hip joint is the source of pain. It is also sometimes used therapeutically for temporary pain relief for weeks to months.

Chapter 4 - Nonoperative Treatment of Hip Arthritis

Conservative treatment is initially indicated for nearly all patients with hip arthritis, with surgery being reserved for those patients in whom conservative measures are no longer enough. Most patients have usually progressed through the spectrum of conservative treatment by the time they are referred to an orthopaedic surgeon, but it is important to be aware of the nonoperative options that can often suffice for months or years before pain and disability are significant enough for hip surgery.

Activity Modification

It is important to maintain as much activity and joint motion as possible, but impact activities will aggravate arthritis. Running and jumping will often accelerate cartilage loss from the joint. Using an elevator instead of stairs and avoiding uneven terrain are helpful. However, there is significant evidence that remaining active and keeping the hip moving will prolong its life. Many patients worry that they should give up walking or other low impact activities in order to try to preserve the hips, but a sedentary lifestyle actually will shorten the life of the hips. The key is to focus on low impact activities, such as swimming or cycling. These are the best forms of exercise with arthritic hips as they do not require significant weight bearing across the hip joints. For patients who do not have access to a pool or a stationary bicycle, leisurely walking will also maintain hip range of motion, strength, and function.

Canes

Canes or walking sticks are useful, particularly when the arthritis affects only one side. Some canes have multiple feet or prongs (e.g., a quad cane) to increase stability for patients with poor balance.

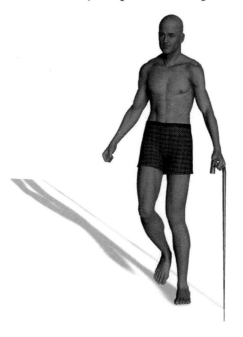

Figure 4-1. Use the cane on the opposite side of the bad (or recently operated on) leg.

A surprising number of patients use canes on the wrong side, however! It is important to use the cane in the opposite hand from the bad hip or knee. This allows you to lean away from the bad leg, taking weight off of it. It is also important to adjust the height of the cane so that the hand height rests comfortably along your side, preventing stooping or poor posture.

Weight Loss

Significant weight loss for obese patients can make a dramatic difference, although in actuality, relatively few patients are successful in losing weight because arthritis limits their ability to exercise.

Weight loss is probably the single most effective intervention the patient can undertake on their own. Increasingly, severely overweight patients (300+ lbs.) are turning to bariatric surgery (e.g., gastric bypass surgery) with promising results, although it remains a serious operation. For the average patient who is somewhat overweight, losing 20 lbs. or more can often at least improve their discomfort and may delay the need for surgery.

Weight loss is also important for increasing the life span of a joint replacement. Surgery is also less risky for patients who are not severely overweight. Although in our practice we do regularly perform joint replacements for patients even over 400 lbs., it is with the thorough understanding that their joint replacements may wear out more quickly, and they are at increased risk for complications with surgery. Surgery usually takes longer and is more challenging for the surgeon when the patient is morbidly obese (body mass index > 35), due to the loss of anatomical landmarks, prolonged exposure and closure time, and need for additional assistants at the time of surgery.

Hip Injections

Injection therapies do not "cure" the underlying problems of arthritis, but can be useful for short term relief (potentially for a few months) and for diagnostic purposes.

Steroid injections into the hip bursa (between the muscle layers on the outside of the hip) are usually quite effective for bursitis, and may be all that is required in combination with physical therapy and anti-inflammatory medications for resolution of a patient's symptoms. These injections do not require live x-ray, and the injection is usually administered over the side of the hip in the location that is most tender to palpation.

Injection of the hip joint is a deeper injection. It typically is quick and only takes a few minutes in the office, but it does require the use of a live x-ray machine (fluoroscopy) to ensure that the injection is placed into the hip joint itself. Most surgeons inject a local anesthetic and steroid mixture (often along with a small amount of contrast, which can be seen on the live x-ray). The anesthetic will often make the hip feel immediately better and for a few hours afterwards, and then the local anesthetic wears off. The steroid component often may take 5 to 7 days to fully take effect.

While injection of the arthritic hip joint itself is not a cure, it does have several very useful roles. It is very useful for diagnostic purposes to help determine where a patient's primary source of pain is originating from. Frequently patients may present with both sciatica and hip arthritis; prior to planning hip replacement, it is useful to see if their pain improves (even for just a short while) by injecting the hip. Another common scenario is that of a patient who knows that he or she needs a hip replacement but is looking for a few months of temporary relief (e.g., they are traveling, or have a daughter's wedding coming up, etc.). Most surgeons try not to use steroid too frequently as it does have some side effects (notably, weakening of the bones and tissues, and rarely, infection), but commonly surgeons will consider injections a few times per year to be acceptable.

A newer injection option on the horizon may be hyaluronate injections. While these are commonly used for knee injections (and discussed at length in that section of this book), they are still considered "off-label" and investigational for hip arthritis at the time of this writing. We have periodically used hyaluronate injections into the hip for some patients with good results, but this treatment is not usually covered by insurance or Medicare at the present time.

Nonsteroidal Anti-inflammatory Drugs (NSAIDs)

This family of medications includes aspirin, ibuprofen, naprosyn, and other non-narcotic medications to decrease inflammation. They remain the mainstay of preoperative management of arthritis pain and are usually most useful in the early years of developing arthritic pain.

Most patients experiment with different over-the-counter NSAIDs before finding the one that seems to work best for them. Older NSAIDs such as aspirin and ibuprofen have been around for many years, and newer drugs in this class called COX II inhibitors, such as celecoxib (Celebrex), valdecoxib (Vioxx – now discontinued), and meloxicam (Mobic) have recently been introduced. Many physicians feel that these are not much different from aspirin and ibuprofen in effectiveness, although these medications have fewer

side effects such as gastrointestinal upset. For this reason these more expensive drugs are usually employed when a patient cannot tolerate traditional over-the-counter NSAIDs, typically because of GI upset. Some of these drugs were in the news a few years ago (notably, Vioxx) because there was some concern about heart problems in a small number of patients. These drugs also require monitoring of liver function if taken for a long period of time.

It is important not to take NSAIDs on an empty stomach, or to use them with blood thinners (such as warfarin) unless directed by a physician. Collectively, these medications are responsible for many cases of GI bleeding and ulcers in elderly patients each year. These medications can interfere with kidney function and may lead to swelling in the legs. These medications can also interfere with some blood pressure medications, and it is important to also check with the physician prescribing the blood pressure medication before taking any of these medications.

Although orthopaedic surgeons may provide an initial prescription for a month or two of NSAIDs, it is usually preferable to obtain these from your family physician over the long term because of the need for monitoring after several months of use. Some of these drugs require liver function tests and other testing after prolonged use.

Glucosamine / Chondroitin Sulfate

Glucosamine chondroitin is a "nutraceutical," essentially a supplement that is often found in the vitamin aisle of the drug store or supermarket. As such, it does not typically have to adhere to the same labeling rules as drugs that are regulated by the FDA, and it is not uncommon to see labels proclaiming that it will "re-grow cartilage!" There is not much evidence that it is likely to do anything so dramatic, although there is compelling evidence that it is relatively safe and works by decreasing inflammation in the joint, making at least some patients feel better. Patients with a shellfish allergy should use caution when taking this, as it may cause an allergic reaction. The typical dosage is about 1500 mg of glucosamine and 1200 mg of chondroitin sulfate daily. Most manufacturers sell the two mixed together in a single pill. It is not uncommon to have to take it for two weeks or more before a significant benefit is seen.

Narcotics ("Pain Killers")

Most hip and knee surgeons feel strongly that these do not have a role in the preoperative management of arthritis, and in our practice, we typically do not prescribe them except after surgery or fracture. Narcotics (such as oxycodone, hydrocodone, oxycontin, etc.) are useful for treating significant pain that is expected to get better in a few weeks. When taken for a long period of time, they can have serious side effects, including addiction, constipation, confusion, and a need for higher levels of narcotics to maintain the same level of pain relief. Additionally, patients who have been on narcotics for any significant time prior to surgery are typically more difficult to keep comfortable after surgery because they have developed a tolerance to opiates (narcotics).

Key Points For This Chapter:

- Activities that aggravate hip arthritis should be avoided if possible

- A walking stick or cane used in the OPPOSITE hand can be helpful

- Weight loss is important both before and after hip replacement; it may significantly prolong the life of the joints

- Injections into the hip joint (steroid or hyaluronate) or bursa (steroid) may provide months of temporary relief, but steroid injections should not be administered too frequently

- Anti-inflammatory medications like aspirin or ibuprofen are helpful, but they must be used responsibly to avoid GI problems, bleeding, kidney or liver problems, or interfere with blood pressure medications. For these reasons, it is best to check with your family doctor if you take them for a long period.

- Glucosamine Chondroitin appears to help with joint inflammation and pain, but often needs to be taken for 2 weeks or more to see a benefit

- Narcotics are best reserved for surgery or broken bones (i.e., problems that are acute and will improve in a few weeks), not for chronic problems like arthritis

Chapter 5 - Total Hip Replacement (Arthroplasty)

Modern total hip replacement was first pioneered by Sir John Charnley in England in the early 1960's. Although previous attempts in the early 20th century included ivory prostheses and other materials, Sir Charnley was the first to essentially develop the modern design that has been the basis for subsequent variations and to produce successful results. Over the past half century, hip replacement has become one of the most successful interventions not just in orthopaedic surgery, but in all of modern medicine. Over 95% of patients have good results (probably closer to 98% in large centers). The outcomes have been steadily improving and the life of the implants increasing over the past several decades.

The basic concept of a total hip replacement (also known as total hip arthroplasty) is to replace the ball and socket joint with an artificial ball and socket. After the joint is replaced, there is no longer **any** arthritis in the joint, because the joint is entirely artificial.

At the time of surgery, the ball (femoral head) and socket (acetabulum) are typically quite worn out. Frequently, the femoral head looks very similar to a head of cauliflower in a very worn out hip, covered with lumpy and bumpy osteophytes and areas of exposed bone where the cartilage has worn away.

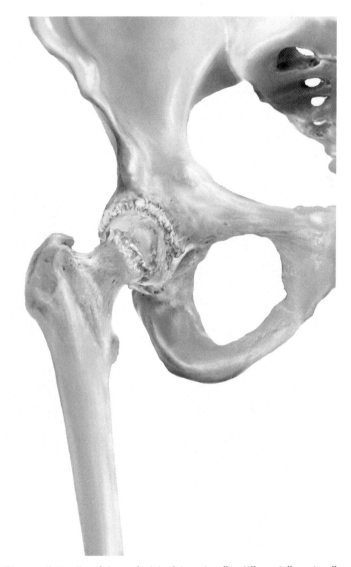

Figure 5-1. In this arthritic hip, the "ball" and "socket" are irregular and covered with spurs (osteophytes). *(Courtesy of Smith & Nephew, reprinted with permission.)*

Nuts & Bolts: The Replacement Procedure

Regardless of the surgical approach used, the same steps have to be performed during the surgery. After exposing the hip joint, the femoral head (the ball of the thigh bone) is cut and removed. The femoral head is usually sent to pathology in most hospitals for routine evaluation, although it is usually kept on the surgical field until the end of surgery in case bone graft is needed.

Next the hip socket (acetabulum) is debrided and scraped clean. Hemispherical reamers are then used to ream the hard, sclerotic arthritic surface of the acetabulum until a bowl-shaped area (similar to the shape the socket is naturally supposed to have) has been reamed out. It is important for the surgeon to ream and prepare this socket at the proper angles; if the cup is placed too steep (vertical pelvic tilt), the hip will have a tendency to dislocate and pop out of the socket. If the cup is placed too flat, the femur will impinge against it when the patient tries to lift the hip out to the side. The optimum pelvic tilt angle is typically about 45 degrees.

It is also important to pay attention to whether the cup faces forward (anterior) or backward (posterior). This is called anteversion or retroversion. Depending on the surgical approach, somewhere between 0 and 15 degrees of anteversion usually is desirable. Too much in either direction, and the hip will dislocate as the leg is rotated inward or outward. Essentially, the acetabular cup (artificial socket) has to be positioned correctly in 2 planes to prevent dislocation or impingement.

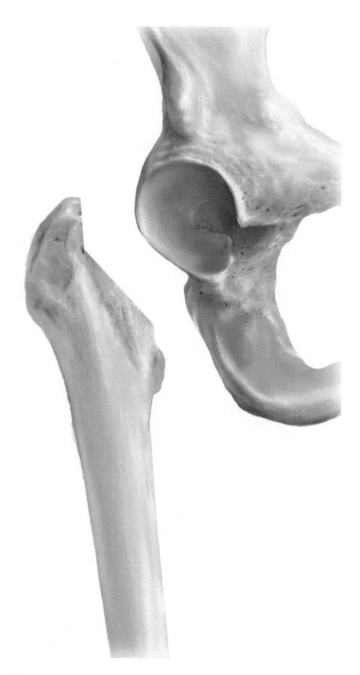

Figure 5-2. The worn out femoral head is resected first in preparation for replacing the ball and socket. *(Courtesy of Smith & Nephew, reprinted with permission.)*

At this point, a hemispherical shell (artificial socket) is then installed in the pelvis. In the early days of hip replacement, this usually entailed cementing a plastic socket into the bone, and this is still done for some special circumstances (such as performing a hip replacement for an elderly patient with a hip fracture, who has very soft bone that may break while impacting a press fit socket into place). More commonly today, however, a porous coated metal shell is impacted into place. The tight fit usually is adequate to hold the shell in place, although sometimes screws are used if supplemental fixation is needed. The back of the metal shell is often coated with a porous metal surface that allows the bone to grow into the prosthesis.

A liner is then inserted into the socket shell. Traditionally, this has often been a plastic liner (polyethylene), and this is still the most commonly used material because of its lower cost. However, in an active and/or young patient, a ceramic liner or even a metal-on-metal liner may be inserted. The choice of bearing surfaces is discussed later in a another chapter.

Sometimes there is not enough of a socket in the pelvis to support the metal shell. This is not an uncommon scenario in complex revision surgeries, traumas, or in cases where the patient has severe hip dysplasia and never formed a proper socket. In these cases, the socket is typically reconstructed with a metal cage that fits over the large hole and may be supplemented with bone graft to fill the defect. Multiple screws anchor the construct into the pelvis, and a liner is then cemented into the cage.

Next attention is returned to the femur. The femoral canal, or the softer marrow within the femur, is then reamed and/or rasped to prepare a slot for the stem. The hardest bone is the outer cortex, like a pipe. Ideally the prosthesis should rest as close as possible to that outer cortex, but if the slot is reamed too much, a fracture can result from the surrounding bone being too thin. Conversely, if not enough of the interior of the femur is reamed, the prosthesis will be sitting in soft bone and may sink (subside) over a period of time and fail.

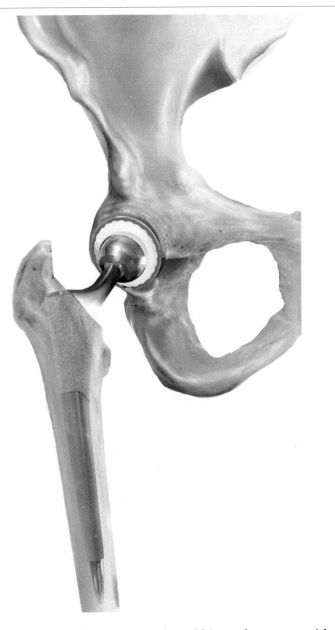

Figure 5-3. The completed total hip replacement, with the new stem and socket solidly implanted within the bone. *(Courtesy of Smith & Nephew, reprinted with permission.)*

As with the acetabulum, some femoral stems are designed to be cemented in place and others are designed with porous coating so that the bone grows into the stem. Increasingly, there has been a movement to use cement less and less and to use porous noncemented stems. These tend to last longer, as the cement surrounding the stem can loosen like grout around a tile over time, leading to failure and a need to revise the hip replacement. Most surgeons agree that noncemented stems last longer in younger, more active patients, and cemented stems have a less desirable track record for these patients. The cement (polymethylmethacrylate) also can have some dangerous effects on blood pressure during anesthesia, and although this side effect is uncommon, it can be serious in elderly patients. However, cemented stems are still used in some places, especially for patients who have very poor bone stock (e.g., very osteoporotic) that is unlikely to adequately grow into the prosthesis.

Trial components are often used at this point during the surgery to determine first if the hip is stable without dislocating, and to determine second if the leg lengths are reapproximated.

In recent years, some newer hip replacements offer modular designs. These designs allow different sizes for the femoral stem, the neck, and the head. In this way, the patient's anatomy can be reapproximated more accurately than with a single piece (i.e., monoblock) prosthesis. This is discussed in further detail in the chapter on implants for those that are interested.

Finally, the femoral head (ball) is selected and impacted onto the top of the femoral stem (trunion). The hip is relocated and carried through a range of motion to test how well it reapproximates normal motions and how stable it is to resisting dislocation.

The incision is then closed in multiple layers, sometimes over a drain if the surgeon feels it is needed based on the degree of bleeding seen from the tissues and if there is a large amount of space (e.g., in an obese patient) that needs to be closed down under suction. The drain typically is removed in one to two days.

Depending on the surgeon and the quality of the skin (i.e., healthy skin, very thin elderly skin, thin skin from being on prednisone for a long time, thick scar tissue from previous surgeries, etc.), various closures

may be used for the skin. Staples are strong and are often used in high tension areas (like the knee). Absorbable sutures leave nothing behind that needs to be removed. Sometimes large nylon or prolene sutures are used for scarred skin that may be difficult to hold together otherwise. A sterile dressing is applied and the patient is then taken to the recovery room.

Hemiarthroplasty (Partial Hip Replacement)

A hemiarthroplasty is a partial hip replacement (just the stem and ball, without replacing the socket). In this type of surgery, the large metal ball fits within the patient's own natural socket. This surgery is most commonly performed on frail hip fracture patients with significant medical risk (because of severe cardiac or pulmonary disease, etc.). This is a faster, shorter surgery that is typically performed to stabilize the hip and allow them to walk, but the surgeon opts not to take the additional operative time (and thus anesthesia risk, bleeding time, etc.) to replace the socket.

Although it is faster (typically by about 25-50%, depending on the surgeon) and has less blood loss, the partial hip will cause wear against the cartilage in the hip socket over time and may require conversion to a total hip replacement at a later date.

Across the country, there is a growing trend to perform total hip replacements instead of the quicker (and less expensive) partial replacements in hip fracture patients who have a significant life expectancy and are active, community ambulators (e.g., they get out and about and are still mobile). Outcomes have thus far been better than partial hip replacements for active patients, although some studies suggest that outcome depends on whether or not the surgeon doing the emergency total hip replacement does total hip replacements routinely. In our practice, if the hip fracture patient walks on his or her own and has his or her mental wits about them (e.g., is not demented and able to follow instructions to prevent dislocation), we will often perform a total hip replacement.

Key Points For This Chapter:

● In total hip replacement (total hip arthroplasty), the ball (femur) and socket (acetabulum) are replaced.

● In a partial hip replacement (hemi-arthroplasty), only the ball is replaced. This is typically for low demand and ill patients with hip fractures.

● There is an increasing trend to use total hip replacements for active patients with hip fractures (femoral neck fractures) to speed recovery and long term function.

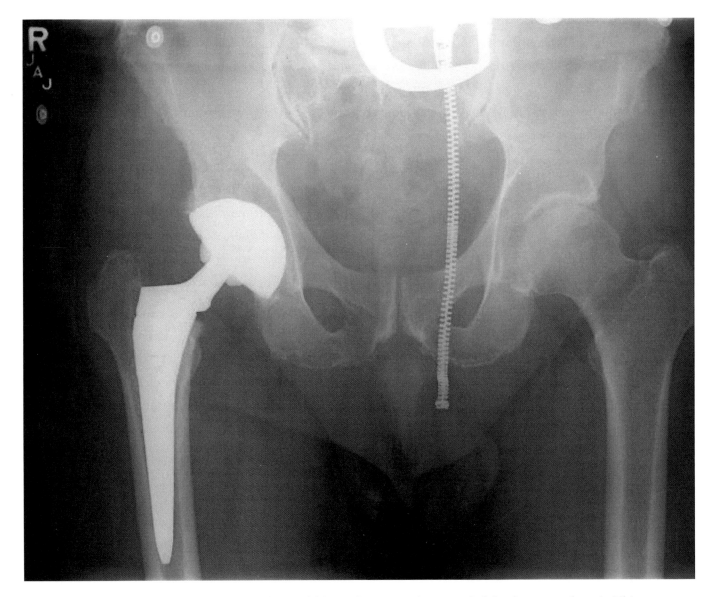

Figure 5-4. Radiographs taken after a right total hip replacement (on your left in the x-ray above). This particular hip replacement design is modular (e.g., assembled in sections).

Chapter 6 - Hip Resurfacing

Hip resurfacing is similar to total hip replacement, except that the top of the femur is capped with a spherical surface (sort of like capping a tooth) rather than cutting off the ball and placing a stem into the femur. Otherwise, the ball moves within an artificial hip socket very similarly to a hip replacement.

Hip resurfacing is not new. A number of the procedures were performed in the early 1980's. However, the engineering and materials were not yet advanced enough, and most designs 20 years ago utilized a metal cap that moved within a plastic liner cemented into the patient's acetabular socket. With these early designs, the cement interface frequently failed and the plastic wore out, leading to early failure in young patients.

The concept has now been revisited, primarily by British surgeons. McMinn and others in Birmingham, England, have developed a newer variation that involves a metal cap that moves smoothly in a highly polished metal socket (both parts are made of cobalt chrome). This device is known as a Birmingham Hip Resurfacing arthroplasty, and while there are some competing designs likely to obtain approval soon, at the time of this writing, this is the first of the newer generation hip resurfacing device approved by the FDA for use in America. It was approved in 2006 and has been been adopted here as American surgeons learn the surgical technique. After visiting surgeons in England, Dr. John Keggi and I together performed the first Birmingham hip resurfacing (BHR) in Connecticut in November 2006, and within a few months, the surgery has now become more widely available by other orthopaedic surgeons as well.

Who Is A Candidate For Hip Resurfacing?

This type of surgery has several distinct advantages and disadvantages when compared to total hip replacement. For the right patient, it is an excellent alternative to total hip replacement, and overseas there is now 10 year follow-up with over 60,000 patients, so far with spectacular results. However, not all of the patients who arrive in my office requesting the procedure are good candidates for it.

The procedure is principally designed for younger, more active patients who need a greater range of motion than typical total hip replacement patients. It is also for patients who need to be able to eventually participate in impact activities such as running. Activities such as hiking, swimming, cycling, doubles tennis, and golfing can all be accomplished just fine with a total hip replacement, but for patients who want to continue to participate in jogging, downhill skiing, martial arts, or other rigorous activities this represents the best option available.

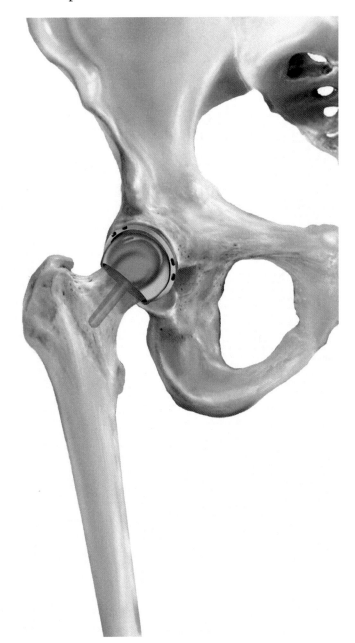

Figure 6-1. A Birmingham Hip Resurfacing. A metal cap is affixed the top of the femoral neck and moves within a metal socket. (*Courtesy of Smith & Nephew, reprinted with permission.*)

The FDA currently recommends that candidates be 65 years or younger for men, and 60 years or younger for women. This is primarily because of the increased bone density needed to support the cap, and it is possible to fracture the bone at the neck of the femur . Some patients outside of this age range can be considered, although a bone density scan may be needed to determine that bone quality is adequate. Similarly, some younger patients may not be candidates if they have soft or osteoporotic bone.

Because of the metal on metal bearing surface and subsequent accumulation of metal ions in the body, patients with true metal allergies or kidney disease are also not good candidates (the kidneys are responsible for excreting metal ions). For this reason, we also will typically avoid women of childbearing age who may potentially become pregnant (a ceramic total hip is a better choice for those patients).

Finally, because the resurfacing relies on the bone of the femoral neck and head to support the metal cap, patients who have significantly abnormal bone anatomy are not good candidates. This may include those with previous fractures or surgeries, or patients who have such advanced degenerative disease that there is insufficient bone stock to support a resurfacing. These patients are better served with total hip replacements.

Advantages of Hip Resurfacing

There are several advantages to hip resurfacing. The first is increased range of motion. Because the diameter of the ball (femoral head) within the socket (acetabulum) is the largest size possible, the allowable range of motion is much better than that of total hip replacement and begins to approach that of a normal hip.

The metal on metal surface also lasts a very long time with very minimal wear. The bearing itself will last many years; the wear rate is much better than that of metal on plastic (polyethylene) and approaches that of ceramics. Typically, the bone it is attached to wears out (loosening) before the implant fails.

The metal on metal surface is also durable enough to allow impact activities, which should generally be avoided with ceramic hips.

Another advantage is that when a resurfacing does eventually wear out, it can be converted to a normal total hip replacement relatively easily. In comparison, most total hip replacements require more complex revisions (to remove a stemmed component from the femur) and revision prostheses. The resurfacing "cap" is simply removed similarly to a normal femoral head when converting to a total hip replacement.

Disadvantages of Hip Resurfacing

As I frequently explain to patients who arrive for evaluation for hip resurfacing, there are some disadvantages that they need to be aware of. As noted previously, one disadvantage is the metal on metal wear and the accumulation of metal ions in the body. Although there is a large amount of data from over 60,000 Birmingham hip resurfacings worldwide, as well as data on other metal on metal hip devices going back at least 40 years, there has not been any decisive indication that the metal ion accumulation is responsible for any detrimental problems in most patients. However, we may yet find in the decades ahead that there is a small increased chance of problems from heavy metal ion accumulation. This would also present a problem for the (very rare) patient who has a true metal allergy.

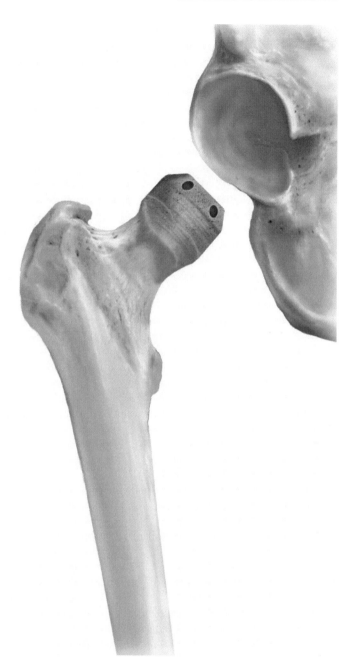

Figure 6-2. In contrast to a total hip replacement, the femoral head is milled down and retained in a resurfacing. For this reason, the bone quality of the femoral neck must be adequate to avoid fracture. *(Courtesy of Smith & Nephew, reprinted with permission.)*

Another disadvantage is that there is some risk of the femur breaking. The neck of the femur, or the part just below the ball, can fracture either during or after the procedure if it is not strong enough. This risk is low, however.

One more disadvantage is that although the resurfacing is less invasive to the bone, and patients retain more "factory original" parts, most surgeries to perform hip resurfacing actually are more invasive than traditional total hip replacements. This is because the femoral neck and "ball" are cut away with total hip replacement, leaving room to work on the socket (acetabulum). If the bone of the femoral neck is to be spared for hip resurfacing, this necessitates extensive soft tissue releases to get the femoral head out of the way so that the socket may be worked on.

The Future of Hip Resurfacing

In most of the world, hip resurfacing is conducted via a posterior approach that requires splitting gluteus maximus and completely detaching gluteus medius, piriformis, quadratus, obturator, and gemelli muscles from the bone in addition to releasing the entire hip capsule circumferentially. This often results in a pronounced limp until the muscles have healed sufficiently.

We recently began performing anterior Birmingham Hip Resurfacings (e.g., from the front) in 2007 after working on adapting the approach at the cadaver laboratory at Yale and working on the necessary modifications for 6 months. At the time of this writing, our practice has been able to offer this surgery with a direct anterior approach that does not require detaching any of the above muscles, only releasing the capsule, because the approach is from the front of the hip. This better preserves the blood supply to the femoral head as well. Early results have been very promising with this new procedure and approach, and thus far patients have had faster rehabilitation and few significant complications. It remains a very active area of research and will be followed in the years ahead.

Key Points For This Chapter:

- Hip resurfacing arthroplasty includes replacing the socket (acetabulum) like a total hip replacement, but uses a cap on the top of the femur instead of replacing the ball with an artificial femoral stem and ball

- Earlier attempts in the 1980's did not last because of the engineering and materials used. Modern designs use metal on metal bearings instead of plastic sockets.

- Current techniques (e.g., Birmingham Hip Resurfacing) are new in the U.S., but have 10+ year track records in Europe and Australia with very good results.

- There are distinct advantages (range of motion, impact activities) and disadvantages (metal ion accumulation).

- The surgery is recommended for patients 60-65 years and younger with good bone density and high activity demands that cannot be adequately met with traditional total hip replacement.

- The surgery is NOT recommended for patients of any age with significant osteoporosis of the hip, kidney disease, metal allergy or sensitivity, altered femoral anatomy (e.g., prior surgery or trauma), or [in our practice] women who may become pregnant.

- Posterior BHR is actually more invasive than a routine total hip replacement, requiring splitting of the gluteus maximus and complete detachment of several muscles from the bone (gluteus medius, piriformis, obturator, gemelli, quadratus).

- Anterior BHR is very new, and thus far has been offered after adapting the approach to the surgery at the anatomy laboratory at Yale with cadaveric studies. However, patients so far who have had the anterior approach for hip resurfacing have rehabilitated much faster because no muscle detachments are required.

Chapter 7 - Surgical Approaches For Total Hip Replacement And Resurfacing

Hip replacements are carried out through one of several different surgical approaches. Over the decades, surgical approaches have been developed that go in through the front of the hip (anterior), between the front and side of the hip (anterolateral), the side of the hip (lateral, or transtrochanteric), and through the buttock (posterior). There are advantages and disadvantages to each, and there is a great deal of controversy among hip surgeons as to which is the "best." All surgeons have a favored surgical approach, and while there are often spirited debates at academic conferences and meetings, it is a testament to the success of the procedure that all of them generally produce good results.

Posterior Surgical Approach

The posterior approach, or Southern approach, is the most commonly used surgical approach for hip replacements in the United States today, although as more interest has been generated in recent years in minimally invasive techniques other approaches are increasingly being used. The patient is positioned on his or her side for this surgery, in what is called the lateral decubitus position.

This approach uses a large, curved incision centered over the buttock. It is usually the largest incision of all of the surgical approaches for any given patient, and requires splitting of the gluteus maximus muscle. The short external rotator muscles are completely detached from the femur, and the hip is dislocated. The femur is twisted around to the front of the patient and rotated inward expose the socket (acetabulum) and femur.

This surgical approach has the advantage of a very large exposure and visualization, but the disadvantage of significant muscle disruption. There is also a higher risk of blood clots because of twisting the vessels.

Some surgeons have recently been utilizing smaller incisions for the posterior approach, often using instruments designed to allow less surgical dissection, but the interval and muscles involved remain the same.

It is more difficult to perform bilateral (e.g., both right and left) hip replacements at the same time with this approach, as it requires repositioning during surgery and placing the patient on the freshly operated side. (In contrast, with an anterior approach, both hips may be replaced more easily during the same surgery, if necessary.) Many patients also note that the posterior incision is on the cheek of the buttocks and may be irritated by sitting.

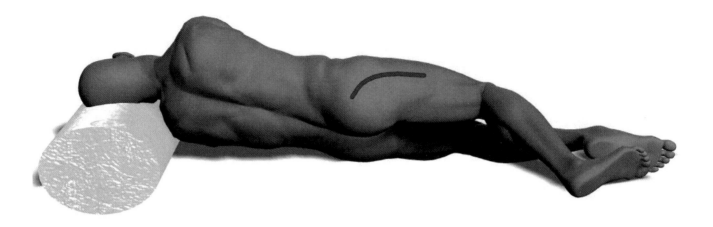

Figure 7-1. In the posterior approach to the hip (the dark line over the buttocks), the patient is positioned on his side (lateral decubitus position). This approach transects the gluteus maximus and detaches several muscles, but allows wide exposure and visualization.

Anterolateral Approach

This approach, also known as a Watson-Jones approach, typically uses a straight incision over the side of the hip, with the patient positioned on his side in a similar fashion as the posterior approach. The surgical approach goes straight down to the femur, but it does require stripping of the gluteus medius muscle from the femur to expose the hip joint. From there, it usually is not necessary to detach the short external rotator muscles, but the remainder of the procedure is similar to the other surgical approaches.

The anterolateral approach is thought by many surgeons to afford a lower dislocation rate than the posterior approach, but a frequent criticism of the approach is that many patients limp for a prolonged period of time while the muscles heal (gluteus medius and gluteus minimus).

Transtrochanteric Approach

When Sir Charnley first began doing hip replacements, he utilized this approach to enter the hip from the side, cut a portion of the femur away to expose the hip joint (trochanteric osteotomy), and then wire the bone back together with the muscles still attached at the end of the case. The approach is very similar to the anterolateral approach except that it involves cutting a portion of the bone (the osteotomy). However, it fell out of favor over the past several decades because of problems associated with re-attaching the section of cut bone. It is mentioned for historical interest here, given that it is not commonly used any longer in most places.

Anterior Surgical Approach

This is the surgical approach that we use in our practice. It involves making one, two, or on occasion (for very large patients, usually 300 to 450 lbs.) three smaller incisions over the front of the thigh.

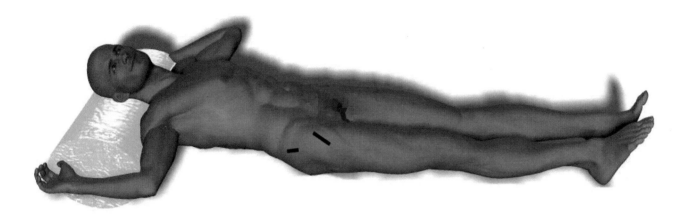

Figure 7-2. The 2 incision anterior approach typically uses one or two smaller incisions over the front of the thigh with the patient supine (laying flat on his back). This also facilitates bilateral surgeries (working on both the right and left sides). The lower incision is used to remove the femoral head (ball) and replace the acetabulum (hip socket). The upper incision, if needed for a large or muscular patient, is used to place the stem in the femur.

The original anterior surgical approach, known as a Smith-Peterson approach, has been around for many decades. In the 1970's, Dr. Kristaps Keggi first developed and published the modified anterior approach that we utilize, making it one of the newest surgical approaches (while orthopaedic implant technology changes all the time, surgical approaches have changed very little in the past 100 years). This surgical approach has been taught to all Yale orthopaedic residents for nearly 30 years now.

In the past 5 to 10 years, there has been increased interest in the U.S. in the anterior surgical approach because of increased patient (and surgeon) interest in minimally invasive surgery. It provides the least disruptive surgical approach, but it is one of the more technically demanding approaches from a surgeon's viewpoint because of the need for increased awareness of the local anatomy and less visualization / exposure with the smaller incisions.

The incision can be either straight or curved, depending on the size of the patient's thigh, and is carried down to the tensor fascia. This fascia is split, and the interval between tensor fascia lata and the sartorius muscle, and then between the rectus femoris and gluteus medius muscles, is opened without having to cut across any muscles. This same approach can be extended proximally (towards the head) and distally (towards the foot) for revision surgeries and even total femur replacements (which are only done rarely, replacing the entire femur and both the hip and knee joints).

For large or muscular patients, a second, smaller incision (usually about an inch in length) is often made over the side of the thigh to pass the stem into the femoral canal so that a larger, single incision is not necessary.

There are a number of advantages to this surgical approach. Intuitively, it makes sense that there is significantly less muscle disruption in approaching the hip from the front and avoiding splitting the gluteal muscles in the buttocks. In fact, the approach only splits the tensor fascia, and then exploits a natural interval down to the hip joint itself, preserving muscle attachments. In contrast, the posterior approach still used in the majority of hip replacements today requires extensive muscle dissection through the gluteus maximus and complete detachment of the short external rotator muscles (piriformis, obturator externus, gemelli, quadratus).

Incidentally, with the anterior approach the incisions are typically significantly smaller and cosmetic, although it is the muscle dissection underneath the skin that is far more important in minimally invasive techniques.

Some surgeons have advocated using live x-ray (fluoroscopy) during the surgery with minimally invasive approaches, particularly this anterior approach. While it is certainly an option for a surgeon who is new to the technique or has any questions about positioning, we do not typically find the additional radiation and x-ray exposure to be justified in most routine cases. Our practice performs hundreds of replacements annually without using x-rays during surgery, but it is mentioned here because you may come across surgeons in some locations that advocate its use.

Some surgeons advocate computer navigation during surgery for the same reasons, although its utility has been debatable in many studies so far, and it is unclear if the benefits outweigh the drawbacks and increased operating time.

The anterior approach is performed with the patient laying flat (supine) on his or her back. This is important for several reasons; there is a lower incidence of blood clots because the hip is not twisted 120 degrees at odd angles as it is in some posterior approaches. It is simpler to match up the lengths of the legs when they are both straight rather than having

the patient on his side for a posterior approach, and the surgeon can easily check that the patellae (knee caps) are even. This position also facilitates bilateral hip replacements, which we perform often, by only having to position and drape the patient once. In contrast, posterior approaches require repositioning and draping, and moreover, the newly operated incision is on the downside against the table while the opposite hip replacement is performed.

The anterior approach avoids the sciatic nerve that runs along the back of the hip, which is the most frequently injured with posterior approaches and can result in a foot drop post-operatively. However, a skin nerve in the front of the thigh (the lateral femoral cutaneous nerve) is at increased risk with the anterior approach, and may rarely result in a patch of numbness over the front of the thigh.

Commentary

In our review of over 2000 anterior total hip replacements by Dr. Kristaps Keggi, published in 2004, the direct anterior approach had a very low complication rate and excellent, rapid rehabilitation. In my opinion and that of many other orthopaedic surgeons specializing in joint replacement, this approach affords the most rapid rehabilitation available, although to be honest there are proponents of other surgical approaches who would dispute that claim. However, there is little disagreement that it does involve the least dissection of muscles.

The most common criticism of the direct single and two incision anterior approach is that it is technically challenging, and for this reason it is not often used by surgeons who do not specialize in hip surgery. Over 50% of all hip replacements in the U.S.

are performed by community orthopaedic surgeons who perform one joint replacement a month or less, and in these situations it makes sense that a larger, posterior approach with better exposure and visualization would be used. However, if the same surgical approach is used three to six times per day, for hundreds of surgeries per year, it becomes easier to see why good results can be achieved by those who use it frequently. Not surprisingly, published studies in recent years have shown that outcomes are better and complications are fewer when total joint replacements are performed by surgeons who specialize in joint replacements and do the surgery more often, regardless of the surgical approach used.

Conclusion – Multiple Surgical Approaches Exist

In summary, there are multiple surgical approaches for hip surgery, and there are also multiple surgeons who advocate one particular approach over others. At our center, we do a great deal of research and publication regarding minimally invasive surgical techniques and feel strongly that the anterior (Keggi) approach has a strongly proven record of superior outcomes over the past three decades, but there are proponents of all surgical approaches at various centers.

If you have a preference for a particular surgical approach, it is in your best interests to look for a surgeon who uses it routinely rather than try to talk your surgeon into a surgical approach that he does not use often. In the end, the best advice for the patient is to find a surgeon whom you like and feel comfortable

with, be sure that he has good surgical outcomes and a significant volume of hip replacements (preferably multiple hip replacements each week!), and allow the surgeon to use the surgical approach and technique that he is most accustomed to.

Key Points For This Chapter:

- There are multiple surgical approaches to the hip, and all get the job done with a high degree of success.

- There is disagreement among hip surgeons as to the "best" approach.

- We use the anterior (Keggi) hip approach developed and described in the 1970's. This approach has been taught to Yale residents for the past 30 years because it does not disrupt the muscles like other approaches, can be used in the supine (flat) position, does not put the sciatic nerve at risk, has a low incidence of blood clots and dislocations, and offers the advantages of minimally invasive surgery with rapid rehabilitation.

- Some surgeons may advocate using intraoperative x-rays (fluoroscopy) and/or computer navigation

- Ultimately, all surgeons use the surgical approach that they personally get the best results with. Studies have shown that the best predictor of patient outcome may be the volume of hip replacement surgeries the surgeon does each month.

Chapter 8 - Surgical Alternatives To Hip Replacement or Resurfacing

The vast majority of patients reading this book to learn about hip surgery will be considering either total hip replacement or hip resurfacing. However, there are some other surgical alternatives that exist. Most of these options were developed in the years before joint replacement or resurfacing was widely available, but some are still performed today in very select cases.

Hip Fusion (Arthrodesis)

In the years before hip replacement, a common surgery for a severely arthritic hip was to fuse the femur to the pelvis with a large plate and screws, effectively eliminating the joint and creating a single bone from the pelvis to the knee.

This surgery persisted even after the development of hip replacements, primarily as an option for young patients (such as laborers) who would otherwise wear out an artificial joint very quickly. Modern designs and materials have mostly made this surgery obsolete, however, and it is very rarely considered today. Few patients in the U.S. would be willing to accept the limitations of a fused hip in the modern era of hip replacement and resurfacing surgeries.

The principal disadvantage of a fusion is that there is no longer any motion at the hip, given that the bones are fused together. This leads to an awkward gait pattern. Sitting and walking are severely affected. Also, the back and the knee typically begin to develop arthritis from "double duty" trying to accommodate the lost motion. The patient needs to have a normal opposite hip and good knees to consider hip fusion.

However, once the two bones have fully grown together, a fusion will rarely need any further medical treatment. There is no implant to wear out, break, or become infected. It is also much cheaper than using a hip replacement prosthesis. For this reason, this surgery is still used in poorer parts of the world where hip replacement is not an option for patients.

Hip Excisional Arthroplasty (Girdlestone Procedure)

If there is a severe problem with the hip joint, such as arthritis or infection, another surgery developed in the early days of orthopaedics was to simply remove the femoral head (or ball). This was called a Girdlestone procedure, and it is still used today for last-ditch efforts at fixing complex problems, typically infection.

Patients can still walk without the femoral head. The weight is borne on the remaining femur, which usually rides against the rim of the socket (acetabulum). It usually does require wearing a substantial shoe lift, however, to make up for the loss of the ball, which can frequently be several inches or more. It is also an uncomfortable gait compared to hips that have had replacement or resurfacing surgeries.

This procedure is used most commonly today for a hip replacement that has had severe complications and cannot be reconstructed or reimplanted. This might be recommended for an elderly or very ill patient who would not do well with a complex revision surgery. Another scenario is a total hip replacement that has become infected, and the infection cannot be cleared by other, less drastic means.

A similar scenario is the patient with a history of intravenous drug abuse (such as heroin). These patients have a high likelihood of infecting the hip replacement if they continue to use I.V. drugs, and many surgeons would opt for a Girdlestone procedure in treating a hip replacement that has become infected in this way.

Hip Osteotomy

An osteotomy means cutting the bone and re-aligning it to heal in a different position or angle. Several types of osteotomies have been used over the last 100 years for the treatment of arthritis and other hip problems.

One particular application that is still used is pelvic and/or femoral osteotomies for young patients with hip dysplasia. If the bone has not formed correctly, it is sometimes possible to cut it and re-align it, such as changing the angle of the hip socket (acetabulum) for a very shallow hip, so that it does not dislocate. This is still commonly used for pediatric patients instead of hip replacement.

Another application for osteotomy is to cut the femur to rotate the femoral head (or ball) so that the worn out or arthritic portion is not in contact as much, and a healthier area of cartilage is used for the weightbearing portion instead. Most osteotomies take a long time (months) to heal because they are surgically created fractures, typically with a long period of limited or nonweightbearing..

Options For Avascular Necrosis (Osteonecrosis)

Numerous operations have been described for treating avascular necrosis. Various rates of success have been reported for each, and these success rates vary widely. Ultimately, most joint replacement surgeons use hip replacement or resurfacing when the femoral head is sufficiently diseased, and the majority of patients do very well.

Most of the operations other than hip replacement or resurfacing are designed to try to prevent progression of the dying bone. A common technique involves core decompression. This essentially means drilling a hole from the side of the hip up into the ball of the femoral head, with the goal of restarting normal bone formation. This is a very quick operation (typically less than half an hour) that usually leaves only a half inch incision. Success depends on how far along the disease is, but sometimes it can stop or even reverse the process of bone death. Most joint surgeons believe that this success rate is probably only about 50% at best, although some claim success rates of 80% with this operation. The rate of success appears to be at least somewhat dependent on the stage of the disease and how much of the bone is involved.

The major downside to the procedure is that it requires prolonged nonweightbearing (or minimal weightbearing), usually 6 weeks or so, with crutches or a walker. This is not usually an option for patients who have both hips affected. Additionally, there is some risk of a fracture because of the long tunnel drilled through the upper femur.

Another variant of a core decompression involves drilling the hole and then placing something inside that hole to support the bone while it heals. Some surgeons have advocated harvesting a 6 inch segment of the fibula from the leg, and placing it into the drilled out hole in the hip. This is called a free fibula grafting, and it usually requires two surgical teams and a surgical microscope to re-attach the arteries to the bone graft. Most joint surgeons are not advocates of this procedure because the results are not as predictable as joint replacements, it is a long surgery with potential complications, and many patients have chronic problems with swelling and pain in the leg where the fibula has been harvested from.

Still another variant of this surgery involves placing a cylindrical core of trabecular metal (very porous metal) into the drilled hole. There is not much data or surgical experience with this yet, however.

Other procedures have been described over the years that directly place bone graft into the area of dying bone in the femoral head. These procedures are sometimes called "light bulb" or "trapdoor" grafting procedures, based on how the femoral head is scraped out and packed with graft (either from below or above, respectively). These surgeries also are typically more involved than hip replacement or resurfacing and often do not have as predictable results or outcomes.

In our practice, we may on occasion recommend trying a core decompression procedure for a young or otherwise very active patient who only has avascular necrosis in one hip, with the understanding that the chances for success may be 50% or less. More often, however, we will perform total hip replacement or hip resurfacing with good results.

Hip Arthroscopy

It is possible to insert a camera and small instruments into the hip joint and have a look around, possibly removing bone fragments or pieces of torn cartilage. This is not a good operation for a patient with significant arthritis, but may on occasion be useful for a patient with bone or tissue fragments (e.g., labral tears) that are causing locking or other mechanical problems in an otherwise healthy joint.

Although this only involves very small incisions to place the camera and instruments through, there are some risks. Nerve injury can occur both from the placement of the instruments as well as (more commonly) from the pressure of the post placed between the legs for counter-traction as the leg is distracted to open up the hip joint. This can cause numbness or loss of sensation in the genital region, which can be problematic and may not return.

If surgery for these types of problems is needed, in our practice we may often recommend an arthrotomy instead, which involves making a small incision over the hip joint and opening the joint in order to fully visualize structures. It is not as minimally invasive, but the incision is still quite small and offers better visualization and less risk to nerves from positioning.

Additionally, arthrotomy is significantly faster than hip arthroscopy, and the surgery can often be accomplished in less time than it takes to complete the complex traction set-up required for hip arthroscopy.

Key Points For This Chapter:

- Hip fusion (arthrodesis) is the process of fusing the femur to the pelvis, eliminating the painful joint but also eliminating motion. It is not used much anymore in this country, although it is an alternative still used mostly in poorer countries, especially for young laborers.

- Hip excisional arthroplasty (Girdlestone procedure) involves removing the ball from the hip joint. Patients can still walk on the femur, although it is very shortened and does not function as well as a joint. This is most commonly used for treating severe medical problems or complications such as intractable infection.

- Hip osteotomies (cutting and re-aligning the bone of the pelvis, femur, or both) are a potential alternative to joint replacement, especially for very young patients and children, but require a long time to heal and are often not the best choice for adults.

- Numerous alternatives to hip replacement/resurfacing exist to try to slow or stop the progression of avascular necrosis before it reaches the point that replacement or resurfacing is needed, but none have very high rates of success.

- Core decompressions and other procedures for early avascular necrosis do not have the same predictability and success rates as joint replacement, but may be recommended because of patient age or other factors.

- Hip arthroscopy involves looking inside the hip joint with a camera and working with instruments through very small incisions, typically used for removing a loose body or a labral tear. It is not recommended for patients who already have arthritis. There are some risks with this surgery, and an arthrotomy (or opening of the joint) may be recommended instead for some patients.

Chapter 9 - Hip Prosthesis Designs

Since Sir Charnley performed the first modern hip replacements in the 1960's, there have been continual advancements in the evolution of the hip replacement prostheses themselves. Some of these changes have involved the materials used for the femoral stems and cups, the materials used for the ball bearing surfaces, whether or not cement is utilized, and most recently, modular designs that allow different components to be put together at the time of surgery to "custom build" a prosthesis to the right length and offset.

Decision To Select The Implant

There are many factors that the surgeon must consider when choosing an implant. First, there are many different options today, including cemented versus noncemented components, modular versus monoblock (single piece) designs, and several different choices for bearing surfaces. Second, there are advantages and disadvantages to each of these choices, and the rest of this chapter presents the most common pros and cons of each design choice. Third, it is a fact of life that cost is increasingly becoming a factor in this aspect of the surgery, and in many places around the country, surgeons are at least somewhat limited by hospitals (and in some regions, limited quite severely)

to what choices of implants can be considered. In some cases, the newest and best prostheses may cost more than the hospital will be reimbursed by insurance or Medicare, and the hospital may have policies in place to limit the use of such implants.

In general, operating rooms may often allow orthopaedic surgeons who perform a large volume of joint replacement surgeries more latitude because of the numbers of patients in which the hospital does not take losses. Surgeons performing a small number of joint replacements each year (once a month or less) may not have that flexibility at the community hospital.

Cemented And Noncemented Hips

In the early days, all hip replacements were secured into the bone with cement. This is still the case for most knee replacements, for mechanical reasons that will be discussed in the knee section. The cement itself has changed very little, and polymethylmethacrylate (PMMA) cement is utilized in all sorts of orthopaedic applications where cement may be needed. In actuality, it functions more as a grout than true cement, filling in the porous spaces around a prosthesis.

The cement offers the advantage of initial strength. At the time the patient leaves the operating room, it is as strong as it will ever be. This is also the downside, given that cemented components eventually loosen much like a cobblestone in a cemented walkway. Although cement was initially used for securing all components into bone, over the years noncemented designs and porous coatings have evolved that outperform cemented replacements in active patients with stronger bones. However, there are still surgeons who prefer and advocate cemented hip replacements, and nearly all surgeons will on occasion use cement for very poor quality bone, such as performing a partial hip replacement for a hip fracture in an elderly patient with very osteoporotic bone.

Another advantage of cement is the the ability to mix antibiotics into it, like vancomycin or gentamicin. These antibiotics leach out over a period of months. Sometimes antibiotic-impregnated cement beads or spacers are placed temporarily into a joint in order to fight deep infection for weeks or months.

Besides concerns over mechanical loosening, however, cement can greatly complicate revision surgeries in which the interior of the bone has to be scraped or drilled extensively to remove old cement. The cement itself can also cause serious complications during surgery as it sets, sometimes dropping blood pressure dramatically. While not usually a problem, this can be of significant concern to both the surgeon and anesthesiologist if it occurs during surgery.

Noncemented components use porous surfaces that are designed to allow bone to grow into the metal. These provide a very strong interface after a period of months. Once the prosthesis is solidly fixed, it is quite secure. Some implants are coated with hydroxyappetite, a chemical that further promotes bone formation.

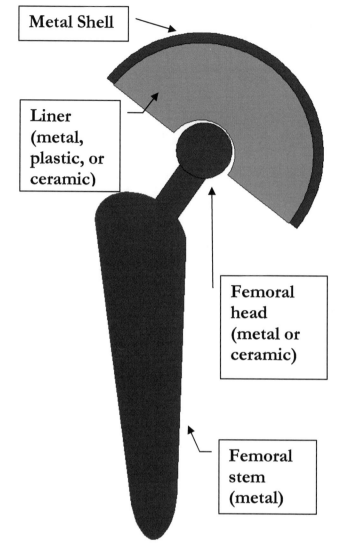

Figure 9-1. The parts of a typical total hip prosthesis. There is usually a metal acetabular shell (socket), a liner, femoral head (ball), & stem.

Introduction to Bearing Surfaces

The ball and socket comprise the bearing surface. The bearing surface can be made from a different material than the stem and socket components. This allows the surgeon to have some flexibility in using the best material for the stem and socket (such as titanium, which is structurally strong but does not work well as a bearing) and the best material for the bearing itself (such as cobalt chrome or ceramics).

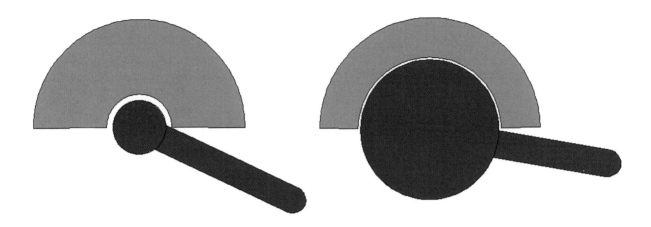

Figure 9-2. Stability vs. wear. Larger diameter bearings have a better range of motion (and therefore more stability and less risk of dislocation), but the increased contact surface area causes increased bearing wear.

There are a number of different materials now used for the bearings, but the most common are metal on plastic (e.g., a cobalt chrome ball on a polyethylene liner), metal on metal (both the ball and the liner are made from cobalt chrome), ceramic on ceramic (both the ball and the liner are made out of a very tough, low-friction ceramic material such as alumina oxide), or hybrid materials (such as cobalt chrome ball with a zirconium oxide surface against a plastic liner). There are distinct advantages and disadvantages to each.

Stability vs. Wear in Bearing Selection

One principal concept is the trade-off between stability and wear. A larger diameter ball is more stable and has a greater range of motion before the stem impinges against the side of the socket. This makes it more difficult to dislocate and improves a patient's range of motion. However, as the diameter of the ball increases, the surface area of the bearing increases as well, and the bearing wears out more quickly.

For many years, the optimum bearing size was thought to be 28 mm, providing a good balance between stability and wear against a plastic (polyethylene) liner. However, as better materials were engineered – particularly ceramic and metal on metal hip designs – it has become feasible to use larger bearings to improve range of motion.

Bearing Surfaces – Cobalt Chrome on Polyethylene (Metal on Plastic)

The initial ball bearing surfaces utilized metal balls (of different types) on plastic liners, which is still common today in most places. Some early designs in the 1960's and 1970's tried industrial coating materials like Teflon, which seemed like a good idea, but turned out not to work very well as the Teflon wore away very quickly.

The most common type of bearing surface in use today is a cobalt chrome ball (femoral head) that moves against a polyethylene (plastic) liner inside the hip socket. This is also the least expensive type of material that is available. It does not typically last as long as ceramic on ceramic or metal on metal ball bearings, but works fine for a patient with a life expectancy of 15 years or so. It may be used for younger patients because of some of its other advantages (see below), but it is usually explained to them that they will likely need revision surgery in 10 to 20 years to at least replace the liner as the plastic portion wears away.

Advantages of metal on plastic bearings include the low cost, essentially zero risk of fracture (as compared to ceramics), minimal metal ion accumulation (as opposed to metal or metal bearings), and relatively rare occurrence of problems with clicking or audible noises. Polyethylene liners also can come with raised lips on one side or constrained designs, which help to prevent dislocation at the cost of range of motion in a hip in which the metal socket is imperfectly aligned or when the patient may have trouble following instructions (e.g., early dementia or noncompliance).

The principal disadvantage of metal on plastic is that the metal head wears away the plastic liner over time. Small particles of plastic accumulate within the joint, and the white blood cells in the body try to swallow and dissolve these particles. However, when the white blood cells are unsuccessful at digesting these artificial particles, the white blood cells may burst, releasing the chemicals (enzymes) used to dissolve foreign bodies and bacteria into the bone around the prosthesis. Over time, this causes large cysts to form and areas of loosening occur around the prosthesis, called osteolysis, that frequently leads to a revision surgery at some point as the hip replacement becomes loose.

In the 1990's there were some new discoveries that led to better plastic liners. The sterilization process used then included gamma irradiation in air (bathing the components in radiation to sterilize them). After some time, it was discovered that this process caused the plastic to become brittle after a few years (the reason was because air molecules integrated into the plastic), causing some plastic liners sterilized in this fashion to not last as long as liners used prior to or after that time period. As a result, major changes were made in the industry, and since the late 1990's plastic (polyethylene) liners have been manufactured differently in order to provide a significantly longer lifespan. But it will likely be another 10 or 15 years before we truly know how much longer these materials last than the 10-15 year lifespan of previous plastic liners.

Bearing Surfaces - Metal on Metal

A metal bearing on a metal surface will never break, and it will wear out much more slowly than metal on a plastic liner, usually lasting for decades. Because of the slow volumetric wear rate, a larger diameter ball can be used than with metal on plastic liners, which gives greater stability and range of motion.

Additionally, the wear particles that do form in a metal on metal surface are very small metallic particles, which are much smaller than the plastic particles and do not cause as much of the process of osteolysis like the plastic particles do.

However, we do know that metal on metal bearings generate higher levels of heavy metal ions in the body. These accumulate in the lymph nodes, liver, spleen, and other places in the body. There has been at least the theoretical concern that high levels of these heavy metal ions might cause health problems later, possibly even some types of cancers. However, this debate continues today after many years of such concern, and most surgeons agree that there is little evidence so far of detrimental problems from this in the tens of thousands of patients who have received such implants.

The metal ion accumulation is the reason that most surgeons recommend against metal on metal surfaces for patients with kidney disease (because it is chiefly responsible for eliminating metals from the bloodstream) or in women who may potentially still have children.

We caution our patients about these potential concerns, but balancing this risk for younger, active patients is the known risks that occur with large revision surgery in 10 to 15 years that will be more likely with metal on plastic bearings. Some types of implants, such as hip resurfacing implants, are only available in metal on metal bearings because these are the only materials that will hold up in the design over time. (Ceramic resurfacings still face engineering problems for use in resurfacings because the socket component would have to be made out of all ceramic and risk fracturing with the thin dimensions required.)

Ceramic on Ceramic

Ceramics have been around in hip replacement applications for about 30 years, but they have only recently begun to receive popularity in the U.S. over the past 10 years.

These ceramics are high performance, highly polished bearings that are usually made from alumina oxide or zirconium oxide. They have a very long life, with the lowest friction and wear of any of the bearings. Additionally, what little wear particles are generated are usually biologically inert (think of sand), and do not cause problems with metal ion accumulation or osteolysis as is seen with plastics.

However, there are some disadvantages to ceramics. They can very rarely fracture, although the reported rates for the current generation of ceramics is one in thousands. Most of the fractures that we have seen have occurred as the result of a trauma that was significant enough to have broken a bone or because of design problems in which the stem of the femur impinges against the ceramic liner over time. Ceramic bearings can also sometimes cause audible noises (probably about 1 in 400), such as squeaking or clicking.

Ceramics are also very expensive compared to metal on plastic, and many hospitals across the country may force their surgeons to use the less expensive implants when the hospital would receive less payment for the implant costs (typically this happens with Medicare or Medicaid, which reimburses the hospitals – and physicians – only a fraction of what private insurance or payers will cover). Increased public awareness and pressure for Medicare and Medicaid to cover the cost of these more expensive implants is needed if those patients want to have access to the higher performance implants.

Hybrid Materials (Oxinium™)

Some newer materials have been available in recent years that offer unique characteristics. One such material that we frequently use is Oxinium™, which is essentially a metal (cobalt chrome) head that has a ceramic surface (zirconium oxide). This is a patented material developed by one of the larger orthopaedic implant companies, Smith & Nephew. The metal head undergoes a special treatment process to form a ceramic layer, so that the general idea is that the final bearing surface has the benefits of both ceramics and cobalt chrome: the ceramic surface has very low friction and wear without metal ion accumulation, and the cobalt chrome ball cannot fracture as ceramics rarely can.

This is a relatively new, expensive material, and it has been in widespread use since the late 1990's. The data so far after the first decade of use are promising, and the laboratory studies show that these materials last a very long time in simulators. In our practice, we typically will select this material for someone who needs better wear characteristics than are seen with cobalt chrome on polyethylene but may not (for various reasons) be a good candidate for a ceramic on ceramic bearing surface. It is also useful for patients with a metal sensitivity or allergy, since it generates very few metal particles or ions as the bearing wears.

Oxinium^{TM} is of particular interest in knee replacements, as is discussed in that section. Ceramic bearings are not usually a feasible option in most knee replacements, and we have found this material to be very useful in knee replacement prostheses.

Modular Hip Replacement Designs

A newer development in recent years has been the movement towards implants that have interchangeable, modular parts rather than a single piece design that comes in several sizes. Some modular designs allow different sized components for the femoral stem, the femoral neck, and the femoral head (ball) in order to achieve hundreds or even thousands of different combinations to try to achieve the best reconstruction possible. In this way, the surgeon can increase the offset without making the leg longer, change the angle that the device sits within the socket (anteversion or retroversion), or reconstruct a difficult and distorted joint such as a dysplastic hip. Smaller incisions can also often be used as the prosthesis is assembled in place, somewhat like a ship in a bottle.

A potential downside to modular components is the question of whether or not they may fail at the modular junctions, whereas a single piece design may be less likely to break. However, modular designs have been continually evolving, and it appears likely that more and more hip replacements in the future will be modular to allow greater flexibility in reconstructing patients' individual anatomy.

Other Factors

This chapter is by no means an exhaustive list of the different factors used in implant selection, but it acquaints you with the major design differences and the general benefits and drawbacks of each.

The surgeon also considers a number of other factors in implant selection as well. The geometry of the femur varies considerably from patient to patient; some patients have a narrow "champagne glass" shape to the femoral canal, and others have a straight cylindrical shape. There are prostheses that fit each of these and other geometries. Some patients have either a valgus or varus femoral neck, meaning that there is an excessively large or small angle to the neck-shaft relationship and implant changes have to be made accordingly.

Not infrequently, one leg is already longer or shorter than the other, and the surgeon needs to plan on how to try to correct that problem at the time of surgery if possible. Some femurs have a significant bow to them or even old fractures, and these require additional strategies to make the artificial joint fit down the canal (such as cutting and re-aligning the femur to make it straighter).

The acetabular socket can be misshapen or too shallow (as in hip dysplasia) and will require additional reconstruction to solidly accept the metal cup. Some sockets are too deep (called acetabular protrusio), and reaming the socket will likely result in a hole through the pelvis that will need to be dealt with structurally. Some sockets have very poor bone and need an implant that allows for extra screws to be used for supplemental fixation. Some patients have large cysts present on their x-rays that may be large enough to plan on grafting and filling at the time of surgery.

Some bones have cysts or thinned areas that will need to be examined and possibly grafted. If areas of bone are significantly weakened, strut grafts need to be wired or cabled around the shaft for additional support.

Some patients have already had previous surgeries or old fractures, which can greatly add to the complexity of the case if the surgeon needs to plan for old scar tissue and the removal of old hardware such as pins, screws, or plates.

Some implants are cast, and others are forged. Some stems are made of titanium and others may be made from cobalt chrome or other alloys. Some stems have a very low modulus of elasticity, which means they do not bend very much, and can have mechanical implications. Some stems have a clothespin design at the bottom to allow for greater flexibility in the implant and less thigh pain. Some designs are too stiff or too solid, resulting in stress shielding, a process which causes bone to absorb and disappear over time because it is not adequately loaded enough.

In the end, there is a long list of factors that can play a role in the implant selection, but your surgeon is trained to consider the pros and cons of each and uses these factors to reach a decision for which implant to use.

Key Points For This Chapter:

- Cemented prostheses use bone cement to fix the prosthesis to the bone. These are useful for patients with very poor quality bone. Cement is not used as much today for hip replacements as in the past, because the cement can loosen (especially in younger patients), make revision surgeries more difficult, and cause intraoperative problems with blood pressure.

- Noncemented prostheses are often porous coated so that the bone grows into the prosthesis. This provides a strong bone interface without the need for cement, but does require stronger / better quality bone.

- Metal on plastic (polyethylene) bearings are still the most commonly used and are the least expensive. They do not fracture and do not cause high levels of metal ions, but they wear out the fastest. The plastic particles can lead to early loosening of the hip replacement (osteolysis).

- Metal on metal bearings are durable and do not fracture, and they last a very long time with minimal wear, but do produce higher levels of heavy metal ions in the body. These are used for hip resurfacings.

- Ceramic on ceramic bearings last a very long time with minimal wear and produce biologically inert wear particles, but can fracture on rare occasions, sometimes make audible noises, and are very expensive.

- Oxinium™ is a hybrid material that has a zirconium surface over a metal (cobalt chrome) ball, with some of the advantages of both ceramics and metals.

- Modular hip replacements are a newer development that allows greater flexibility by building the prosthesis in place with different femoral stems, necks, and heads.

- Many other factors, including the geometry and quality of the bone, are used in determining the optimum prosthesis for the individual patient.

Chapter 10 - Hospitalization For Hip Replacement or Resurfacing

Typical hospitalization times usually range from three to four days for most total hip replacement surgeries and hip resurfacings, with younger patients often needing less time in the hospital. The events leading up to and surrounding the surgery are similar for all joint replacement surgeries and are detailed in a later chapter, including information about preoperative testing, blood donation, anticoagulation, etc.

In general, most of the patients treated in our practice can expect a general timeline similar to the one presented here, although there certainly can be variability depending on other medical issues and the exact type of surgery performed.

Day of Surgery

Most patients arrive at the hospital early on the morning of total hip replacement or resurfacing surgeries. This process is detailed in a later chapter because the process is the same for both hip and knee replacement surgeries. The surgery usually takes about 60 to 90 minutes for routine primary (first time) surgeries, although it certainly can take longer if the patient has had prior surgery, is very muscular or obese, is having more than one procedure performed, etc.

The time in the operating room for most hip surgeries is actually longer than the hour to hour and a half needed for the procedure, because the anesthesiologist needs time to perform the spinal or general anesthesia, the patient needs to be positioned and prepped with antiseptic scrub, and sterile drapes have to be placed.

Most patients are in the recovery room for two or three hours after the surgery, at which time routine x-rays are taken to check on the implants and surrounding bones. Patients also need to be fully awake and have a stable blood pressure before being transferred to their hospital room upstairs. However, most hip replacement and resurfacing surgery patients can actually get up with assistance that first evening.

First Day After Surgery

Physical therapy starts in earnest on the second day of hospitalization. Most patients are allowed to fully bear weight on the affected hip(s), but it is important to work with the physical therapist to ensure that there is no dizziness from medications or anesthesia. We typically allow most patients to fully bear weight right away if the postoperative x-rays show everything to be in good position and there are no special circumstances.

Some surgeons will routinely limit weightbearing for noncemented prostheses, with the idea being that the bone needs time to adequately grow into the porous coating. However, many surgeons (as in our practice) would prefer to start full weightbearing as soon as possible by "press-fitting" the prosthesis tightly at the time of surgery. Your surgeon and physical therapist will let you know your weightbearing status after surgery.

Most patients use a walker at first, and when they are ready and steady enough, they progress to a cane. There is a great deal of variability in how long the process takes, because everyone has different levels of physical stamina, balance, muscle strength, etc. As a general rule, younger and more active patients and thinner patients will graduate to using a cane quicker than their older or heavier counterparts.

Most IV lines, urinary catheters (if you have one – not all patients require this), and surgical drains (again, many patients do not have any drains, depending on the size of the patient and surgical factors) are removed on the first or second day after surgery. Although many patients may be apprehensive about removal of drains or catheters, most are somewhat surprised to find that this is not usually as uncomfortable as they expected and literally takes only seconds.

Second And Third Days After Surgery

There are typically daily blood tests to monitor hemoglobin levels and metabolic parameters, and patients are monitored to make sure that they do not have any complications after surgery such as a blood clot or a serious hematoma (more on this later). Blood

thinners of some sort (either aspirin, heparin / enoxaprin injections, or warfarin) are usually started the day after surgery also (the exact regimen depends on the patient, surgeon, and medical factors).

Total hip replacement patients have a list of precautions to follow so that the hip does not dislocate in the initial weeks while the soft tissues are healing. The exact precautions depend on the surgical approach used, but generally it is advisable not to flex the hip beyond 90° (or level with the pelvis while sitting), cross the legs, or pivot on a planted foot on the affected side. Anterior hip resurfacing patients, on the other hand, do not have these precautions because it is very difficult to dislocate those prostheses. The physical therapist works with you to practice your recommended exercise regimen and learn what you should and should not do after leaving the hospital.

After three or four days in the hospital, most patients are ready to graduate either to home with visiting nurses/physical therapy or to a short term rehabilitation facility. This depends on how well the patient is able to get around on their own and their age, but most often the need for a rehabilitation center is dependent on social factors. If a patient lives alone or does not have adequate help, or if the living arrangements cannot be changed to accommodate staying on one level, then a short stay at a rehabilitation center is more likely.

Discharge

The first post-operative visit in our practice is usually between 3 and 6 weeks after surgery. If there are staples or non-absorbable sutures, these are usually removed by the visiting nurse at 2 weeks after the surgery. Most patients undergoing joint replacement surgery in our practice today have absorbable sutures with a type of surgical "glue" for the skin, which does not require any suture/staple removal.

Patients are discharged with a detailed list of instructions on what to do, what not to do, and what to call about (such as fevers, wound breakdown, calf swelling, etc.). These are described in greater detail in a later chapter.

If patients are going to a rehabilitation facility, then instructions and detailed dictations on the hospital course are provided to the facility. If they are going home, then the discharge planner usually coordinates for visiting nurses, home physical therapy, and any necessary equipment (such as a raised toilet seat if needed, walkers, etc.).

Key Points For This Chapter:

● **Most primary (first time) hip replacements and resurfacings require 3 or 4 days in the hospital.**

● **Most first time surgeries take between 60 and 90 minutes per side, depending on patient size and other factors.**

● **Patients usually arrive the day of surgery unless they have a medical problem that necessitates earlier admission (such as blood thinners that cannot be discontinued).**

● **Patients are usually up and walking within 24 hours of the surgery.**

● **Each day, mobilizing and getting up are the most important factors. It gets progressively easier.**

● **Most patients go home with home visiting nurses/physical therapy, but some go to rehabilitation facilities depending on age, general physical condition, and social factors.**

Chapter 11 - Home Life And Exercises After Hip Replacement or Resurfacing

It is very important to continue with physical therapy and exercises after any joint replacement or resurfacing procedure. Once patients are out of the hospital, the surgery may be finished but the physical therapy and rehabilitation are just beginning. Mobility and strength steadily improve with each passing day.

This may sound daunting to someone who is contemplating surgery and currently dealing with a painful hip, but in actuality, most patients find that the pain they have after surgery is quite different from the preoperative pain. Patients commonly remark immediately after surgery that they no longer feel the grinding, deep joint pain with weightbearing that they had previously, and that most of the discomfort after surgery is muscular pain in the area of the incision. Perhaps even more importantly, this type of discomfort steadily resolves and actually improves the more patients are up and about using their new hip.

Getting Home And Transportation

Most patients are able to go home in a regular car, and within a few days, can certainly use the car as a passenger to get to the hospital or physician's offices. Generally it works best to use a vehicle that allows you to stretch your legs out in front of you, and you should avoid any small cars or cars that are very low to the ground as this may require flexing the hips beyond 90 degrees.

Patients often ask about other transportation options. Generally, most patients are able to travel in a regular car, but those going to a rehabilitation facility may prefer use of a wheelchair car or ambulance. These services can be arranged by the discharge planner at the hospital, although most insurance companies and Medicare do not cover such transportation costs.

Generally it is best to avoid nonessential travel out of the house for about 7 to 10 days after a total hip replacement. Most of our patients are able to go for short rides or to a restaurant after about a week. Hip resurfacings can expect to mobilize somewhat quicker. Although some younger and more active patients have actually returned to office (desk) jobs for short periods after the first week, it is typically best to do the exercises / physical therapy and otherwise rest in the first few days after surgery. We generally recommend planning on taking 6 weeks off from work, and up to 10 weeks for a very physical job.

You should not take any extended car trips for 5 weeks. This is primarily because of the prolonged sitting and the increased risk of blood clots.

Driving is usually not recommended until 2 or 3 weeks after discharge, if you have good control of your right leg, and if you do not have any other medical conditions that prevent you from driving. If you have other conditions besides your hip that may impede your driving (such as low blood pressure or neurologic issues), check with your family physician before driving. Obviously, if you have lightheadedness or are still taking narcotic medications, then you should not drive.

We typically recommend that patients practice driving in an empty parking lot, such as an empty school or church parking lot on a Saturday. It is also a good idea to take a family member or friend with you, and if you BOTH feel comfortable with your ability to drive, then begin driving short distances and gradually work up to longer trips.

It is important to understand that you have to take legal responsibility for determining when you are safe to drive. If you feel you are unsafe, then wait until you feel more confident.

Getting Around On Your Own Two Feet

You should be walking at least 4 or 5 times per day, increasing your distance each time. **Walking is your most important exercise after a hip replacement or resurfacing.** It will increase your stamina and strength, decrease stiffness, help to prevent blood clots and constipation, and you actually will feel much better if you are mobile.

However, when you are not walking, remember your rest periods in bed with leg elevation. These breaks are important to prevent swelling. Keep the legs elevated above the level of the heart. Flex the ankles up and down whenever you think about it, which promotes circulation. You may walk frequently, but in general you should spend two hours, twice a day, in bed with the legs elevated for as long as there is persistent swelling in the leg. If your leg and calf suddenly become much more swollen, warm to the touch, and painful in the calf, it can be a sign of a blood clot and you should call the surgeon's office.

During the first week at home, you should not sit in a chair for more than 3 times a day for 30 minutes each time (usually at mealtime). After the first week, this can be relaxed if there is not significant swelling or discomfort. Sitting periods can slowly be increased to a normal routine after the first week.

Total Hip Precautions In Our Practice

Do not cross your legs or extend your hip or leg backwards during this time period. Do not cross the midline with your affected leg(s). Do not internally rotate the leg ("pigeon-toeing").

Always sit in a straight back chair (i.e., dining room chair, no couches, no low recliners, etc.) for 6 weeks following surgery.

When you sit down, slide your foot (on your operated side) out in front of you. Do not lean forward when sitting in a chair during the first few weeks, but if you must lean forward, then be sure to spread your knees apart as you do so (which places the hip replacement into a more stable position).

Do not pivot on the operated leg(s). In particular, do not plant the foot on the operated side and turn, leaving the foot planted and rotating on the hip. This can predispose to dislocation.

Figure 11-1. Hip precautions: DO NOT cross the legs over the midline in the first 6 weeks.

Hip replacement patients should not force hip flexion beyond 90-100 degrees for the first 6 weeks. Resurfacing patients who have had an anterior approach (not a posterior approach) typically can range the hip to whatever is comfortable, but should not force their range beyond the limits of comfort.

Stairs

You may begin using stairs as soon as you feel comfortable. Some patients with good stamina and muscle strength may practice stairs at the hospital before going home; others will take a few weeks to build up their strength. The most important factor is to be safe, and always use a handrail for balance as you begin using stairs again. If you feel unsteady, then you may use a sitting position to scoot up or down the stairs.

When going up stairs, lead with the **unoperated** leg, and when coming down, lead with your **operated** leg. (If both legs have been operated on, then you can use whichever leg is more comfortable.)

Showers and Toilets

Different surgeons may have different guidelines, but in our practice, we generally allow showering 2 days after discharge if the wound is dry. Gently towel the area dry after showering. Do not shower or get the wound wet until 2 days after the wound has become completely dry, and do not allow it to get wet if there is still some drainage.

A shower stool is a good idea for the first 6 weeks after surgery. This can be helpful to avoid slipping and falling.

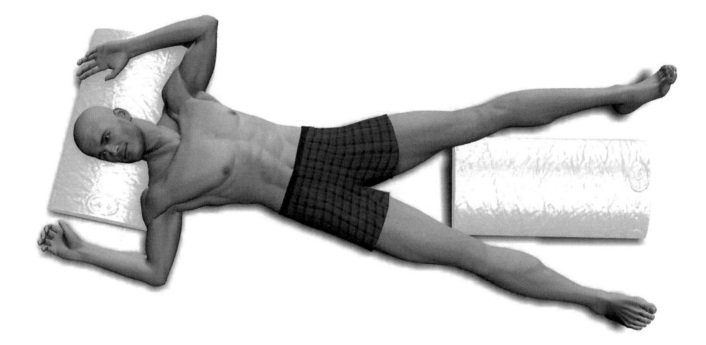

Figure 11-2. Hip Precautions: DO sleep with a pillow between the legs for the first 2 weeks.

Figure 11-3. Hip Precautions: DO NOT sit with the legs crossed over the midline.

Figure 11-4. Hip Precautions: DO sit like this, with the legs apart and in a straight backed position.

Figure 11-5. Hip Precautions: DO NOT flex more than 90° in the first 4 to 6 weeks unless otherwise instructed by your surgeon. This includes bending at the waist beyond 90°.

Do not take tub baths for at least 6 weeks. This is primarily to avoid the motions involved with getting in and out of a tub, but generally it also is not a good idea to completely submerge the surgical site for a couple of weeks.

You may use a regular toilet unless the toilet is unusually low, or unless your surgeon instructs otherwise. For most total hip replacement patients, a raised toilet seat should be used if you find you have to flex your hips above 90 degrees getting on and off of the commode.

Sleeping at Night

Use a pillow between your legs for 2 weeks. Some patients may have more stringent instructions if they have had a dislocation in the past. Try to sleep on your back for the first few weeks to avoid laying on the operated side if possible.

Getting Dressed

Since you are not supposed to cross your legs or force hip flexion beyond 90-100 degrees for the first 6 weeks, putting on shoes and socks will be a challenge at best. Most patients find they need some assistance

An anterior approach surgery has the advantage of having the incision on the front of the hip, and is therefore more comfortable when sitting than when you have had a posterior approach (with the staple or suture line on the buttock). However, some male patients have noted that the anterior incision line may chafe with briefs, and you may find boxers more comfortable.

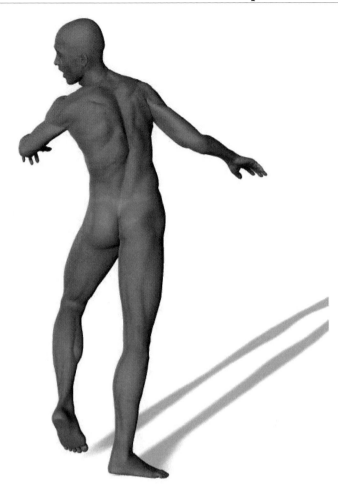

Figure 11-6. Hip Precautions: DO NOT pivot on the operated leg.

Figure 11-7. When going up stairs, go UP with the nonoperated (stronger) leg first.

with this for the first few weeks, although a sock grabber and reacher instrument can be helpful. These are sometimes available from the physical therapist before going home or otherwise can be found at a medical supply store.

Ultimately, when you have healed from your surgery, the best way to practice putting on shoes and socks and clipping your toenails is to slide your foot up along the leg and rest it on the knee, keeping the hip pointed outward at about a 45 degree angle. This is a safe position. Crossing the legs with the knee going over the midline, as ladies frequently sit, is not as ideal and should be avoided for at least 6 weeks.

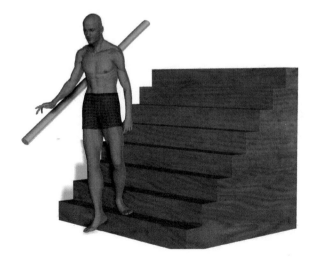

Figure 11-8. When going down stairs, go DOWN with the operated (weaker) leg first.

Exercises

The exercise program after joint replacement is not temporary, but continuous. It is an important part of the ongoing management of your total hip replacement or resurfacing.

As noted above, **walking is the most important exercise.** You should walk at least 4 or 5 times a day, increasing the distance each time. It is better in general to walk for shorter periods with rests in between than to attempt a marathon session once or twice a day. Rest periods are helpful in between.

The walking should be at a slow, steady pace on level ground. (I often recommend going to the mall several weeks after surgery for most patients, as it is level ground and weather is not a concern.) Walking faster will not be particularly beneficial, and if you strain the muscles by walking too quickly, it is possible to have some muscle bleeding and swelling in the first week or so. A slow and steady gait, on the other hand, is very beneficial.

The following exercises are the ones we recommend in our practice, primarily with anterior approach hip surgeries. If you are reading this and have another surgeon, be sure to check with him about your exercise instructions and routine. Your physical therapist, with orders from your surgeon, may also introduce additional exercises to work on specific muscle groups tailored to your needs.

Ten sets of the following exercises should be done each day, and at least 10 repetitions of each exercise should be done during each set. The standing exercises should be done while holding on to a table or using a crutch or cane for balance. If your balance is poor or you feel unsteady, then focus on the exercises that are performed lying down or sitting until you feel more steady on your feet.

Also note that while it is not unusual for exercises to generate some discomfort, significant pain is typically a reason for caution. If a particular exercise is too uncomfortable, then focus on other exercises. If you develop any problems that prevent you from continuing, such as lightheadedness, shortness of breath, or chest pain, then it is best to stop and contact your physical therapist or surgeon.

Figure 11-9. Bend the hip and knee in a standing position; do not flex the hip beyond 90 degrees.

Bend the Knee And Hip

Bend both the knee and hip in a standing position, lifting the leg up and down 10 times. Do not flex the hip beyond 90 degrees (level with the pelvis). Hold on to a table or walker for balance.

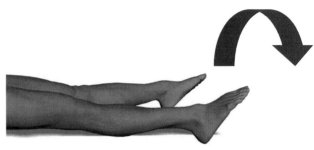

Figure 11-11. Ankle motion in circles.

Ankle Motion

Lying flat (e.g., in bed), move the ankle in a circle. As you get stronger, try to lift the leg up while making the circle motion. Repeat 10 times.

Figure 11-10. Lifting the leg out to the side (known as hip abduction). Use a table, rail, or walker for balance.

Lift the Leg Out To the Side

Stand with the knees straight. Then lift the leg out to the side, hold there for 5 seconds, and then return to standing. Repeat 10 times. Hold on to a table or walker for balance.

Be careful not to pivot on an operative leg if you have had both hips replaced during the same surgery.

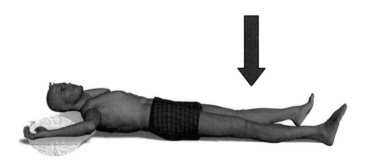

Figure 11-12. Knee isometric strengthening. Try to push the knee downward against the floor.

Gluteal Isometric Strengthening

Lying flat, keep the legs straight and a little apart. Squeeze the muscles of the buttocks together for 5 seconds, compressing the buttock cheeks, then release. Repeat 10 times.

Knee Isometric Strengthening

Lying flat, keep the legs straight and a little apart. Try to tighten the thigh muscles and push the knee downward against the floor or bed, holding the contraction for 5 seconds. Repeat 10 times.

Figure 11-13. Heel Slides. Slide the heel back and forth, while keeping the heel in contact with the bed or floor.

Heel Slides

Lying flat, slide the operated foot up as far as you can while keeping the heel in contact with the floor or bed, then allow it to slowly slide back. Repeat 10 times.

Knee Range of Motion

While laying flat, place a rolled towel or round pillow under the knee, then extend the knee so that the foot rises completely off the floor or bed. Hold it up for 5 seconds, then release. Repeat 10 times.

Figure 11-14. Knee Range of Motion. Try to extend the knee and foot completely off the floor.

Key Points For This Chapter:

- Keep in mind that these are the post-operative recommendations in our practice and that they apply for most but not all patients; always check with <u>your</u> surgeon about your specific limitations and instructions after surgery.

- Walking is the most important exercise after surgery.

- Take frequent breaks and rest periods during the first week.

- No extended car trips for 5 weeks.

- You can start driving in a few weeks if you feel safe to do so. Practice driving in an empty parking lot first! Do not drive while still taking narcotics.

- Shower 2 days after discharge if the wound is dry and your surgeon instructs you that it is permissible to do so. However, no tub baths for at least 6 weeks.

- Go up stairs leading with the nonoperative leg, come down leading with the operative leg.

- Sit in a straight back chair and avoid low couches, recliners, and low toilet seats (use a raised seat for low toilets).

- Your exercise program is not temporary, but continuous. It is an important part of the ongoing management of your joint replacement.

Chapter 12 - Life After Hip Replacement or Resurfacing

Most patients have minimal or no pain by 3 months (or sooner) after hip replacement or resurfacing, and the majority return to our office and report that their discomfort level, activity, and quality of life are all dramatically improved. However, it is not unusual to have occasional muscle aches and persistent (but usually slight) swelling of the thigh and extremity for several months.

Depending on numerous factors, including the surgical approach, amount of surgical work needed, and particularly the state of the musculature around the hip before surgery, some persistent limp is usually expected for a while. In some cases, a limp may be persistent for a long time after surgery (most often for a patient who has had significant muscle atrophy – or wasting – from longstanding disuse of the hip prior to surgery).

Returning To Work

There is a wide variation in how soon patients return to work. It primarily depends on what you do and also on your physical condition before surgery. Obviously, a young person who is in good health aside from a bad hip will be back to work much sooner than a patient who is severely overweight and deconditioned

or who has multiple other medical problems, but most patients get there eventually if their health is reasonable and they work at the rehabilitation.

Some patients return to desk jobs within several weeks. Others who have very physical jobs, such as laborers, may need to take 10 to 12 weeks until they are able to meet the demands of their job.

Essentially, we recommend that everyone returns to work when they can function safely and with reasonable comfort. Each patient is responsible for determining if he or she can safely perform the activities of his or her job (no one else knows what the job entails better than the person who has to perform it!). If there is some accommodation by the workplace (e.g., to allow someone who normally works a standing job to do light duty at a desk for a short while), most patients are back to at least some limited work by 6 weeks, and physically demanding duties typically about a month after that.

Every patient is different, and while surgeons can offer some estimation of the average time recovery will take, it may vary considerably.

Activities after Total Hip Replacement

In general it is best to avoid impact activities after hip replacement. Although some patients engage in activities such as jogging or contact sports, they usually do so against medical advice. It is best to avoid situations where repetitive impacts or sudden jolts might occur. Impacts will decrease the life of the replacement and increase the likelihood of early loosening, possibly necessitating revision surgery.

Low impact activities such as walking, golf, cycling, swimming, hiking, or ballroom dancing are good sources of activity and cardiovascular exercise, and these activities are well tolerated by most joint replacements (assuming your general medical condition allows such activities). Skiing on gentle slopes is usually safe, although we recommend against downhill skiing that involves significant twisting and turning. Most joint replacements will last for many years with proper care and low impact activities.

Some patients enjoy yoga or pilates, and these activities are usually fine for routine exercise with some modification to accommodate the range of motion recommended by your surgeon. However, it is best to avoid awkward positions and hyperflexion of the hips. Certainly, touching the toes to the ground over your head is a risky position and may lead to dislocation! In general, it is usually best to try to keep hip flexion limited to 100 degrees or less after total hip replacement. Some types of hip replacements (e.g., large diameter alternative bearing surfaces, such as metal on metal or ceramics) are designed for greater

range of motion than this, and it is best to ask your surgeon if you have questions about your specific range of motion limitations after surgery.

Sex After Hip Replacement

Another topic that patients are often afraid or embarrassed to ask about is sex after hip replacement (that's why I put it in the book!). Most patients can resume sexual activity 4 to 6 weeks after routine hip replacement if they are otherwise healthy enough.

Generally, the missionary position is safe for both men and women once they feel comfortable enough to resume activities. One area of caution for women is to not hyperflex the hips too far (usually not greater than 100°), but in general, the hips are also abducted (e.g., spread apart) at the same time so that the position tends to be stable.

Straddle position, with the woman on top, tends to be safe for both men and women with hip replacements also (it can be a problem for patients with knee replacements, however, as kneeling is generally uncomfortable).

If you have more creative positions in mind, most things can be accommodated with some common sense and by going slowly. The principal concerns with hip replacements are dislocation by flexing the hips too much or bending over too far to touch the toes. If you have a concern about a particular activity, it is probably best to just ask your surgeon about it rather than do something risky.

Activities After Hip Resurfacing

In contrast to total hip replacements, we do not usually place any specific restrictions on hip resurfacing patients once they are completely healed from the surgery. Although dislocation is possible, it is exceedingly uncommon and difficult. Hip resurfacings can generally tolerate the same types of motion and activity that a native hip can.

One particular area of concern, however, is the possibility of weakened bone in the femoral neck after prolonged inactivity, or particularly in a patient who has had avascular necrosis and possible weakening of the femoral neck. Unlike a total hip replacement, the hip resurfacing keeps the bone of the femoral neck, and if it is substantially weakened there is some reported risk of fracture. In these situations of prolonged inactivity or avascular necrosis, it is best to avoid impact activities for about 1 year after the surgery. Some studies have shown that after a year has passed with restoration of the patient's ability to get around, the bone density in the femoral neck region increases so that weakened bone is no longer as much of a concern. If there is any concern, follow-up bone density scans can be helpful.

Going To The Dentist

In the past, most surgeons have typically recommended the use of prophylactic antibiotics before dental procedures. The reason has been that with dental work, there is often bacteria in the bloodstream for a short while afterwards that (at least theoretically) may lead to infection of the joint replacement. This concept is not limited to prophylaxis after joint replacement surgeries; many cardiologists also recommend antibiotics for patients with heart valve problems for the same reason.

In reality, there are bacteria in the bloodstream on many occasions (such as after brushing your teeth), and the risk of infection of an artificial joint is very low.

There are however some situations in which antibiotics are recommended. The American Dental Association (ADA) and the American Academy of Orthopaedic Surgeons (AAOS) have jointly met and issued some guidelines regarding when antibiotics should and should not be used before dental work. Because this information also applies to knee replacements, there is an appendix with these guidelines at the back of this book.

Longevity of the Implants

Most patients want to have some idea of how long they can expect their hip replacement or resurfacing to last. This is highly variable, however, and there are many factors that contribute to the longevity of the implants used.

Impact activities (running, basketball, and other sports) will increase the likelihood of loosening for a total hip replacement. In addition, most total hip replacements are not designed for these types of sports activities, although hip resurfacings can be. Hip resurfacings can loosen over time, but evidence thus far appears to suggest that they do so less often than hip replacements.

Patient weight has a significant influence over how long the replacements will last. Heavier patients place a larger load on the implants, but conversely, they often are less active (e.g., take fewer steps in a year).

The type of bearing material used factors into longevity. Metal on metal and ceramic on ceramic bearings probably last the longest, followed by hybrid materials, and lastly by traditional metal on plastic bearings. Hip resurfacings are metal on metal bearings and last a long time.

Factors unrelated to the implants may shorten their lifespan, such as infection or trauma. I have treated patients who fell off of ladders or the roof (while cleaning out gutters) who fractured the bone around the implant, necessitating wiring or revision. In general, it usually takes the same amount of energy required to break a bone to damage the replacement, however.

Younger patients wear out their replacements more quickly than older patients. For this reason, many surgeons in the 1980's and 1990's recommended waiting as long as possible before replacing joints in young patients. While we still wait until all conservative (nonoperative) treatments are exhausted, we now recognize that the implants and technology have evolved to the point that even young patients can expect years of use before requiring revision surgery, and hip replacement or resurfacing can now give severely impaired patients their mobility and life back.

Key Points For This Chapter:

- There is wide variation between patients regarding when they return to work (several weeks to several months) depending on how physical their job is and their overall physical condition

- Most patients are driving within 4 to 6 weeks after surgery

- Total hip replacement patients should avoid repetitive impact activities and sports (such as running), although low impact activities should be encouraged (walking, swimming, cycling)

- Total hip replacement patients typically have some range of motion limitations to avoid potential dislocation

- Hip resurfacing patients do not typically have any range of motion restrictions, and additionally, after 1 year may usually resume most sports activities

- Sexual activity can be resumed when patients feel up to it, usually at 4 to 6 weeks.

- Dental visits and antibiotic recommendations are included in the appendix at end of this book.

- Longevity of the replacement depends on many factors, including activity level, patient weight, bearing materials, and external influences (traumas, infections, stroke or neuromuscular disease, etc.)

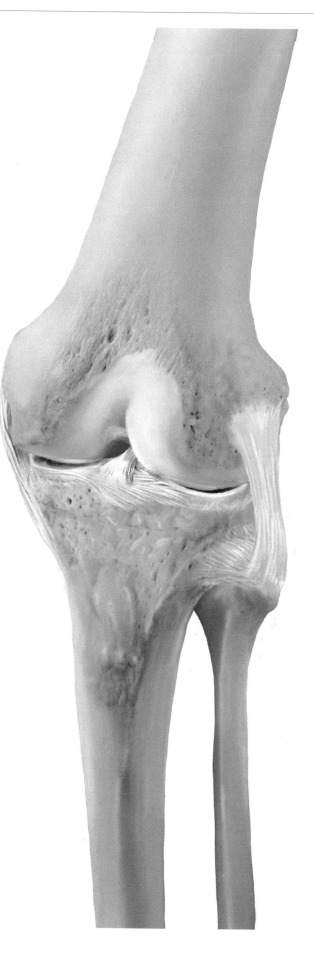

PART II - THE KNEE

Chapter 13 - Knee Anatomy and its Associated Problems

The knee basically functions as a hinge, but in reality it is much more complicated than that. Although we see the hinge motion from the outside, on the inside the motion actually more closely resembles the swing arm of a four-bar linkage. Moreover, there is also a rotational component to knee motion as well. It is a marvelously designed joint but is actually more complex than the hip joint. It may not be surprising, then, that knee replacement did not evolve until after hip replacement.

Bones And Cartilage

The knee joins together the lower end of the femur and the upper end of the tibia (the large bone in the lower leg; the smaller one is the fibula). The patella (knee cap) also is in contact with the femur, where it glides up and down in a shallow trough called the trochlea.

The ends of the bones and the undersurface of the patella are covered with a layer of articular cartilage that is similar to the Teflon coating in a frying pan. The coating is normally about $1/8^{th}$ inch thick, but over time, wear and tear and a number of diseases can lead to the loss of this cartilage coating.

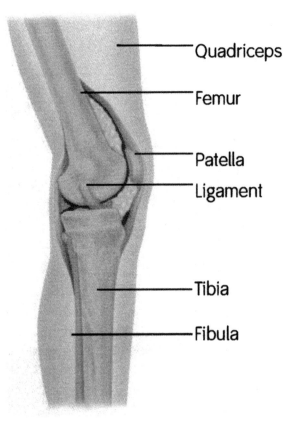

Figure 13-1. Knee anatomy as seen from the side. (*Courtesy of Smith & Nephew, reprinted with permission.*)

When the cartilage is worn away, most patients experience significant and worsening pain from the "bone on bone" contact, and this process is osteoarthritis.

Trauma or injuries can also lead to loss of the articular cartilage. A common mechanism is a "dashboard" type injury in which a bent knee strikes the dash in a car accident, making a "divot" or pothole in the cartilage. This usually does not heal by itself, and over time, the resulting defect can lead to post-traumatic arthritis. Post-traumatic arthritis is also common with any fractures that occur in or close to the knee joint. It is common for patients who experience significant knee trauma to require knee replacement at a later date.

The knee joint is surrounded by a capsule. There is a tough membrane called the retinaculum that covers it along the front, going off either side of the patella. Sometimes this retinaculum can develop tears in an injury, and the patella may not track correctly afterwards as a result.

The Three Compartments Of The Knee

The knee joint has three compartments within it. Two of these are formed where the curved, cam-shaped end of the femur meets the tibia on both the inner (medial) and outer (lateral) side of the knee. The cruciate ligaments separate the medial and lateral compartments and run inside the intercondylar notch. The third compartment is formed by the patella and the trochlear groove of the femur that it glides up and down in (the patellofemoral compartment). There are pockets on the sides of the joint referred to as the medial and lateral gutters, which can sometimes harbor loose bodies and debris.

Knee problems can occur in any and all of these three compartments. Meniscal problems occur in the medial and lateral compartments, where most of the weightbearing also occurs. Problems with the

undersurface of the patella (e.g., chondromalacia patellae) manifest in the patellofemoral compartment.

Sometimes arthritis and other conditions affect only one side of the knee, or affect it to a much greater degree than the rest of the knee, and therefore partial knee replacements and other procedures may focus on just that compartment. At other times, the term tricompartmental degenerative joint disease is used to describe arthritic changes affecting all three compartments, which may dictate a total knee replacement instead of a partial one. This is discussed in greater detail shortly.

Ligaments of The Knee

There are four primary ligaments around the knee that hold it together (there are actually more, but these are the most important ones). The anterior cruciate ligament (ACL) is in the center of the knee and keeps the tibia from sliding forward. It is commonly injured in sports injuries, and in young patients we reconstruct it with surgery. In older or more sedentary patients, an ACL rupture may often be left alone and/or treated with a brace and physical therapy to compensate with better muscle strength. The ACL is usually torn and shredded from chronic arthritic changes in many of the patients who undergo knee replacement.

The posterior cruciate ligament (PCL) is also in the center of the knee and runs opposite to the ACL, preventing the tibia from falling backwards relative to the femur. PCL injuries are not as common as ACL injuries, and the PCL is often intact at the time of knee replacement. In fact, many knee replacement designs make use of the PCL (more on this later), while others replace its function.

On the sides of the knee are the collateral ligaments, both the medial collateral ligament (MCL) and the lateral collateral ligament (LCL). These ligaments are responsible for keeping the knee from giving way from side to side. These are commonly sprained (especially the MCL), but even partial tears often heal without the need for surgery. A hinged brace may be used if the collateral ligaments are torn or injured, which keeps the knee from buckling from side to side.

Menisci

Within the space between the femur and tibia is another cartilage structure called the meniscus. It is a C-shaped gasket made of the same type of cartilage as your ears or nose, and there is such a C shaped segment in each side of the knee (the medial and lateral menisci). It functions as a shock absorber and "chock block" within the knee. It normally does not spread all the way across the knee joint (when it does, this is called a discoid meniscus, which is often problematic and sometimes requires surgery). Instead, it fits around the rim of the knee.

For many years the meniscus was considered "the appendix of the knee," and painful tears were typically treated by removing the entire meniscus, which worked great... for a while. Patients later developed significant arthritis without the stabilizing influence of the meniscus. Today, these tears are treated with arthroscopic surgery instead of complete resection (more on that in a later chapter).

If the meniscus tears, this often leads to pain and clicking in the knee on the affected side. Tears can occur as the result of a twisting injury, or some types simply occur as a wear and tear process, like a rug that has been walked on too many times and develops a frayed spot. If the tear is big enough, the torn flap can flip into the joint space, catching like a door with a piece of carpet stuck in the hinge. This can cause locking, buckling, and catching, and arthroscopic surgery may be needed to correct the problem.

Loose bodies can occur and also cause similar mechanical problems. These can arise from bits of bone or cartilage that break free and rattle about inside the knee joint. At times, these loose bodies can cause pain that may move around in position and cause locking of the knee when they become caught between the bones of the joint.

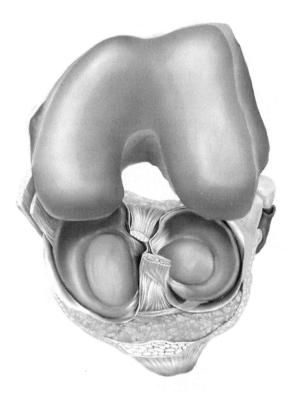

Figure 13-2. Looking down at the knee joint. The two C-shaped structures on either side are the menisci. The cruciate and collateral ligaments are cut in cross-section. *(Courtesy of Smith & Nephew, reprinted with permission.)*

Alignment of The Knee (Varus/Valgus)

The overall mechanical alignment of the knee is important. Although some patients are born with a natural curve in the knee, many arthritis patients develop worsening deformity as one side of the knee wears out faster than the other. If the inner side (medial compartment) wears out first – as is the case for the majority of knee arthritis patients – the knee begins to become progressively more "bow-legged," a deformity known as *varus*. If the outside compartment (lateral compartment) wears out first, then the knee becomes progressively "knock-kneed," a deformity known as *valgus*.

Figure 13-3. Varus knee deformity (bow-legged).

Sometimes the patella does not track correctly in its groove, or may even "jump out" and dislocate frequently. This is usually treated conservatively with a brace and physical therapy, but if it persists there are several surgeries described to change the tracking of the patella.

Side to side deformities like varus and valgus are best assessed with standing x-rays, which show the alignment. The gap normally seen on an x-ray is actually the cartilage, which does not appear on x-ray, and it can be assessed by its absence when the gap disappears or is smaller than it should be with weightbearing x-rays.

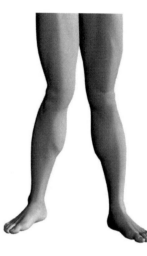

Figure 13-4. Valgus knee deformity (knock-kneed).

Extensor Mechanism

The muscle groups in the thigh (notably the quadriceps) attach to the upper end of the patella, and below the patella is a tough tendon that links the patella to the front of the tibia. This functions like a pulley, and when the muscles in the thigh contract, the tibia is extended and the knee straightens.

Any disruption of the extensor mechanism, either by injury or surgery, can be devastating (more on this in the complications section of the book) – it is like cutting the strings of a marionette, and the knee cannot be extended. Quadriceps tears, patella fractures (with a gap between the fragments), and patella tendon ruptures are all usually repaired surgically because of the importance of this entire assembly in moving the knee backwards and forwards.

Nerves and Blood Vessels

The nerve and blood vessel anatomy around the knee is somewhat simpler than the hip. The major structures (e.g., the popliteal artery and the tibial nerve) run along the very back of the knee, which is usually left alone in most joint replacement surgeries. It is possible to injure the popliteal artery during surgery, but it is exceedingly uncommon (typically seen mostly with complex revision surgery when it does occur). Most surgical approaches to the knee approach from the front of the knee for this reason. There are surgical approaches to the back of the knee (known as the popliteal fossa), but these are usually only used for specific cases such as tumor resection.

There are some arteries that surround the knee known as the geniculate arteries. These may commonly be encountered during surgery and are usually cauterized.

The peroneal nerve passes around the head of the fibula. It is uncommon for this to be injured with surgery, but pressure from a retractor or from swelling and blood within the joint can sometimes injure the nerve, leading to the very uncommon occurrence of a foot drop after surgery. When it does occur, foot-drop usually resolves on its own within months to a year in most cases, but may require use of an orthosis to keep the foot from dragging.

Key Points For This Chapter:

- **The knee is more than a hinge, with complicated motion that is more like a four-bar linkage with an additional rotational movement.**

- **There are 3 compartments to the knee (medial, lateral, patellofemoral), and problems can arise in any and all of the compartments.**

- **There is articular cartilage that coats the surfaces of the bones in all three compartments; it is normally about 1/8th inch thick and wears away with arthritis.**

- **The 2 menisci are C-shaped cartilage gaskets that fit in the medial and lateral compartments. These can develop tears that result in mechanical problems (locking, buckling, catching).**

- **The 4 main ligaments in the knee are the anterior cruciate ligament (ACL), posterior cruciate ligament (PCL), medial collateral ligament (MCL), and lateral collateral ligament (LCL).**

- **The extensor mechanism is formed by the muscles of the thigh (e.g., quadriceps) that attach to the patella, and the patella tendon connects the bottom end of the patella with the tibia. This is responsible for lifting (extending) the knee.**

- **The major nerves and blood vessels are located in the back of the knee; most routine surgeries therefore approach from the front of the knee.**

Chapter 14 - Diseases of The Knee

The last chapter mentioned some of the potential problems that can be associated with specific anatomical features of the knee. Like the hip, some categories of knee problems are broader, and these need specific discussion.

This chapter discusses the most common types of arthritis and other common causes of knee pain, such as meniscal tears, bursitis, osteochondral lesions, and avascular necrosis. Another important cause of knee pain is actually hip joint problems; hip arthritis commonly can present as knee pain. It is not unusual to see a patient who reports they have been told x-rays of his knee are fine, but that he has serious knee pain and limitation, only to find a hip x-ray shows significant pathology in his hip!

Osteoarthritis of the Knee

Osteoarthritis, or the "wear and tear" variety of arthritis, is probably the most common reason for knee pain in mature adults. It is certainly the problem that I see most commonly with patients aged 50 or older who report pain that has been present for longer than several months, although it is by no means the only cause for such pain. It also can be seen in patients of much younger age, especially in obese patients.

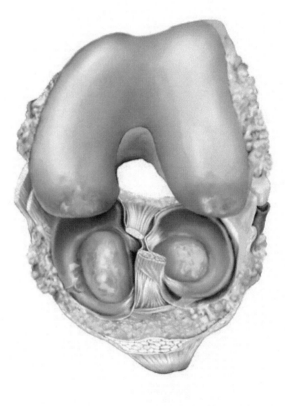

Figure 14-1. Advanced arthritic changes of the knee, viewed from above. Note the worn areas over the joint contact surfaces and the frayed meniscus. (*Courtesy of Smith & Nephew, reprinted with permission.*)

The articular cartilage is a coating of smooth, soft cartilage about 1/8th inch thick covering the ends of the femur, tibia, and patella undersurface. As mentioned in the last chapter, it can wear away with time or injuries, and after skeletal maturity it does not grow back.

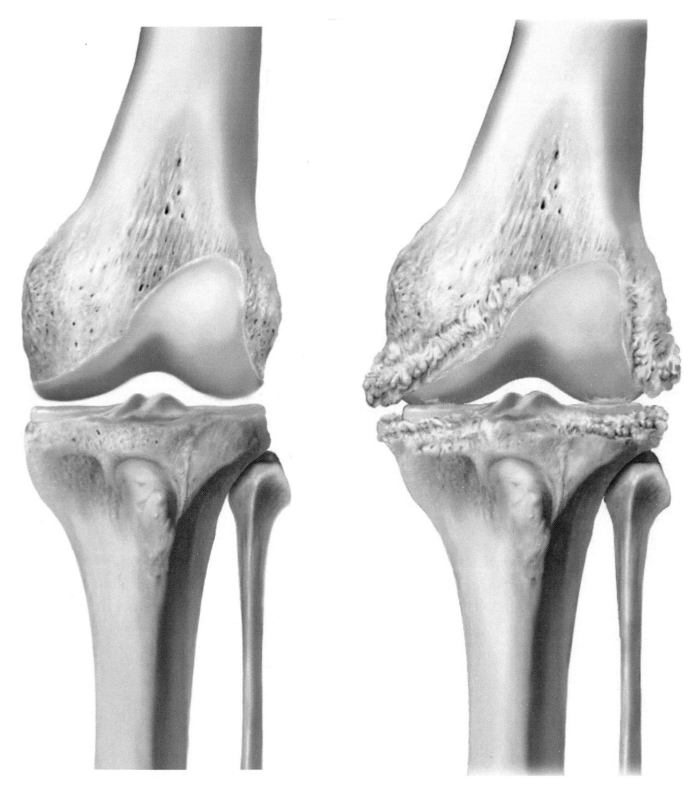

Figure 14-2. Healthy knee on the left, arthritic knee on the right. Notice the eroded cartilage coating and large spurs. *(Courtesy of Smith & Nephew, reprinted with permission.)*

There definitely appears to be a genetic predisposition to developing osteoarthritis.

Frequently, one of the compartments of the knee (usually the medial, or inner side of the knee) may wear out sooner than the other compartments. In this case, some options are available for treatment that are not useful when all three compartments are affected, such as orthotics, off-loader braces, and partial knee replacements. These options for single compartment disease will be discussed more fully in the next couple of chapters.

As the cartilage covering begins to wear away, underlying bone surfaces are eventually exposed (hence the common term "bone on bone" arthritis). As the process continues, the body responds by making osteophytes (large spurs) around the peripheral edges of the joint and often by making more joint fluid. This can result in a large joint effusion, or "water on the knee," which sometimes needs to be drained. The knee will also usually become progressively more stiff as the process continues, and frequently an angular deformity may develop as one side wears out faster than the other. This commonly results in a varus (or "bow-legged") appearance, although sometimes it can result in a valgus (or "knock kneed") deformity. A flexion contracture may eventually develop in which the knee no longer can fully straighten out.

Trauma And Post-Traumatic Arthritis

There are numerous traumatic injuries that can occur in the knee, ranging from contusions and bone bruises to fractures involving the joint surface or nearby shafts of the tibia or femur. Trauma can result in injury to any of the ligaments of the knee or even to multiple ligaments, each requiring treatment that varies

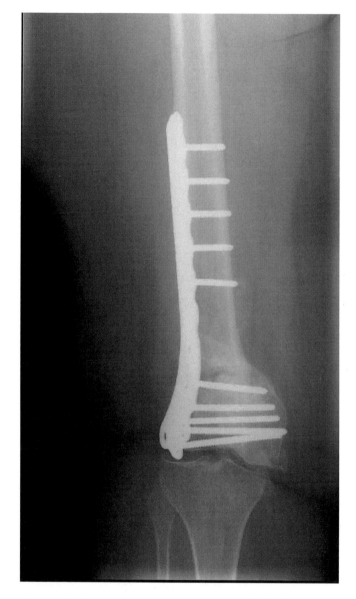

Figure 14-3. Severe post-traumatic arthritis in a patient involved in a motor vehicle accident. Although the bone has healed with the plate and screws, the joint surface of the knee is no longer smooth and is "bone on bone" with large spurs.

from simple bracing and observation (such as a partial medial collateral ligament injury) to extensive reconstruction surgeries (such as an ACL reconstruction). There is a wide variation in treatments of acute injuries, and we will not be able to discuss those in detail here as this book is primarily dedicated to joint replacement and related surgeries.

However, it is important for our discussion here to understand that post-traumatic arthritis can develop after injuries to the knee, and clinically it looks and is treated similar to osteoarthritis. It is not uncommon for patients who have had prior traumatic injuries to the knee to develop post-traumatic arthritis years later. A large percentage of patients who sustain tibial plateau fractures (fractures that involve the weightbearing surface of the tibia) require knee replacement surgeries in the years after the injury, despite adequate bone healing with plates and screws after the initial injury. The smooth surface of the joint becomes disrupted, and over time arthritis results.

For many years, the standard treatment for meniscal tears involved resecting the entire meniscus from the affected side of the knee. However, it turns out that the meniscus serves an important function as a stabilizer for the knee, and without it most people develop severe arthritis in the knee over the years after excision. After the development of arthroscopic surgery, complete meniscectomies became rare, and most surgeons treat nonhealing meniscal tears with arthroscopic surgery in which only the torn portion of the meniscus is removed (and sometimes repaired, if possible). However, it is common to see patients who had complete meniscectomies (or removal of "torn cartilage in the knee") years ago who now require partial or total knee replacement.

Rheumatoid Arthritis

As discussed in the section on hips, another large category of arthritis is that of autoimmune arthritis. These types of arthritis are primarily the result of the body's own immune system attacking the joints, rather than "wear and tear" osteoarthritis, and the most familiar of these is rheumatoid arthritis. Other related "inflammatory" arthropathies include psoriatic arthritis (from psoriasis), lupus, and other rheumatologic diseases.

Frequently patients with these types of arthritis have systemic involvement in which multiple joints are large, swollen, red, and painful. Most frequently the joints of the hands are affected first, but not all patients have that initial presentation. Since newer treatments became available in the last decade or so, we are seeing fewer patients with this type of arthritis who need joint replacement early on.

The synovial lining of the joint often becomes inflamed, red, and boggy with rheumatoid arthritis. Often the inflammation of the lining can be destructive to the joint itself, leading to loss of cartilage and bone over time.

The majority of patients with rheumatoid arthritis (and related arthritic types) are already diagnosed by the time they reach the orthopaedic surgeon, but not always. Blood tests are somewhat helpful but not always definitive in diagnosis of these disorders. Evaluation by a rheumatologist is sometimes needed.

Because of the nature of the disease, tendons and ligaments may rupture over time as the body attacks these tissues. This is why patients with advanced rheumatoid arthritis often have gnarled, deformed hands as the tendons become involved. It is also the reason why many surgeons recommend using a posterior stabilized knee replacement that replaces the posterior cruciate ligament when replacing knees in patients with rheumatoid arthritis, whether or not the patient currently has symptoms with it, because the PCL is prone to rupture in the setting of untreated rheumatoid arthritis.

Osteochondral Defects / Loose Bodies

The articular cartilage surface in the knee, like most joints, resembles a smooth Teflon coating that can wear away over time. Sometimes "potholes" can develop in the surface of the joint, and these can definitely be seen with arthroscopic surgery when the camera is inserted into the joint for a look around. Sometimes a plug of cartilage and underlying bone can break loose, leaving a crater in the normally smooth surface. This crater by itself is harmful, and often leads to accelerated arthritis, but the chunk of bone and cartilage (hence the term osteochondral) can be even more painful and destructive. The loose chunk can become a loose body, moving around inside the knee joint and causing damage, frequently leading to mechanical problems such as catching and locking as the loose piece becomes caught in the knee.

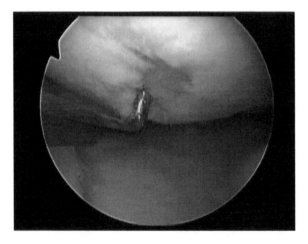

Figure 14-4. Arthroscopic photo of an osteochondral defect. The probe is in the center of a defect - similar to a pothole - in the cartilage surface of the knee.

Chondromalacia

Somewhere along the spectrum between normal joint cartilage and "bone on bone" arthritis in which it is absent, the cartilage goes through a time when it is soft and begins to "delaminate" or peel away. This phase is often referred to as chondromalacia, and it can be seen even in young patients. It does not necessarily mean that serious arthritis is eminent, although it often is a sign of probable degenerative changes that can be expected later.

Surgeons often grade the degree of chondromalacia during arthroscopic surgery, when the joint surfaces can be closely examined with the camera and photographed. These range from minimal cartilage "blistering" to large areas of exposed, denuded bone with the overlying cartilage completely worn away. Consequently, surgeons can often advise patients of the status of the joint after arthroscopic surgery and may forecast the need for knee replacement in the future.

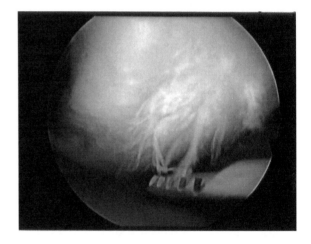

Figure 14-5. Chondromalacia. The fronds being trimmed by the shaver are soft cartilage that is wearing away from the joint surface.

Meniscal Tears

As mentioned in the last chapter on knee anatomy, the menisci are "C"-shaped sections of cartilage (similar to that in the nose and ears) that form gaskets in the knee. There are two of these gaskets, one on each side of the knee. The meniscus on the medial side, or inner knee, is most frequently affected by tears. Tears can

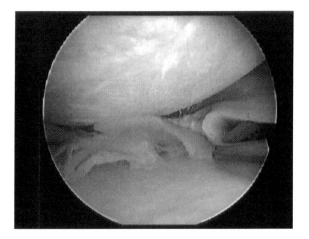

Figure 14-5. An arthroscopic photograph of a large meniscal tear. The femur is on top, the tibia on the bottom, and the hook is probing the shredded meniscus between.

come in a variety of shapes and sizes, ranging from small tears that heal on their own without intervention to large, "bucket-handle" tears that cause the knee to lock, catch, or buckle.

Many meniscal tears result from a twisting injury to the knee, and patients may recall the moment that it occurred and feeling a "pop" as the meniscus tore. Many other patients do not recall the injury at all, often because the pain and swelling may not be problematic until hours or days after the injury. Still other patients develop degenerative tears, in which the fibers begin to fray like a rug that has been walked on too many times, and describe gradual pain and symptoms over time.

Most patients report pain as the principal problem when a meniscal tear develops. Patients with large tears experience mechanical symptoms of internal derangement, such as a catching or popping sensation when the knee flexes and extends. Large tears can cause the knee to painfully lock as the torn fragment becomes trapped in the hinge of the knee. Most patients find that squatting becomes very difficult, and flexing the knee far back (more than 110 degrees)

becomes painful.. The knee may or may not swell, depending on the acuity of the injury.

A Baker's cyst may form in the back of the knee (in the space known as the popliteal fossa), often in relation to meniscal problems. It is common for a Baker's cyst to be identified on an MRI or ultrasound, and patients are often unnecessarily alarmed about the abnormality on their report. Although the pain that patients feel is often in the same area, the cyst itself often is not truly the source of the problem. It is uncommon that cysts are treated surgically, and they often resolve over time after the meniscal tear or underlying pathology has been dealt with (either surgically or by healing on its own). On occasion, the cysts can become quite large, particularly in rheumatoid arthritis patients, and these sometimes are dealt with surgically. Cysts can also rupture, which is usually harmless but may cause a disconcerting sensation of warm water running down the back of the calf.

It is important to note that most patients with significant arthritic changes also have meniscal tears. As the surfaces of the joint become rougher, the cartilage meniscus between becomes torn and ground away. For this reason, orthopaedic surgeons usually do not find an MRI helpful in evaluating a knee that is clearly arthritic on x-rays, as meniscal tears are expected.

Avascular Necrosis (Osteonecrosis)

As discussed in the section of the book on hip diseases, avascular necrosis (or osteonecrosis, as it is sometimes called) is a process in which a portion of the bone around the joint begins to die. This usually occurs because of poor blood flow to the affected area of bone, and the underlying causes for this are many.

Coagulation problems and injury are common factors, but many patients develop the disease for no identifiable reason.

As the area of bone begins to die, the overlying joint surface may collapse, similar to a sinkhole. The process is painful and most patients seek medical treatment before it reaches the point of collapse. The most commonly affected area is the femoral condyle, or the end of the thigh bone, although the disease can also be seen in the upper end of the tibia. If the area collapses, severe degenerative changes will ensue quickly that resemble (and are treated similarly to) severe arthritis.

The process can sometimes resolve on its own before collapse occurs, however, with activity modification and conservative management. Some surgeons advocate limited or no weightbearing using crutches, a walker, or an off-loader brace (which redistributes the load to the opposite side of the knee). Other potential treatments can include arthroscopic drilling, which is sometimes successful in arresting the process.

Infection

Infection of the knee is a relatively uncommon cause of chronic knee pain. It typically presents with an acute infection (sepsis), with purulent material in the knee that necessitates emergency surgery to "wash out" the joint. Patients are usually quite ill with high fevers, inability to move the knee, and severe swelling and pain. It occurs most often in patients who have a compromised immune system (including diabetics) or a reason to have introduced bacteria into the knee joint (such as a penetrating wound or an infected diabetic ulcer lower down the leg).

In unusual cases, a chronic low-grade infection is another potential cause, often from Lyme Disease or similar diseases. A blood test usually reveals infection by these types of bacteria. Aspirating (e.g., drawing fluid out of the knee with a needle) and sending the joint fluid for testing is also helpful in diagnosing infections in the joint.

Knee Bursitis / Tendinitis

Not all from knee pain arises from problems within the joint itself. The tendons around the knee can develop inflammation (e.g., "jumper's knee") and in severe cases the quadriceps or patella tendons can rupture, requiring surgical repair. Tendinitis usually resolves with conservative treatment consisting of activity modification and physical therapy.

There are several bursae around the knee joint as are found around other large joints in the body. The prepatellar and pes anserine bursa are most commonly affected. These fluid-filled sacs help muscle layers slide smoothly over each other, but the sac can sometimes become inflamed and painful. Sometimes it can fill with fluid that needs to be drained. Bursitis often resolves with steroid injections, anti-inflammatory medications, and physical therapy.

Ligamentous Injuries

The four primary ligaments in the knee are the anterior and posterior cruciate ligaments and the medial and lateral collateral ligaments. Basically, each of these ligaments prevents the knee from traveling in one direction (either side to side or backwards and forwards). These ligaments can often be sprained, especially in sports injuries, and these minor injuries usually heal without any intervention beyond bracing and other conservative treatment.

However, some sports injuries can result in complete disruption of one (or occasionally multiple) ligaments. Collateral ligament injuries (the ones on the sides of the knees) may heal with the use of a brace and without the need for surgical repair. Anterior or posterior cruciate ligament (ACL or PCL) injuries are different, however, and complete tears usually do not heal on their own. In fact, simply suturing the ligaments back together (as was tried in the past) usually is not adequate either, and these ligaments usually require reconstruction surgeries using tendons harvested from other body sites (commonly the patella tendon or the hamstrings) or from cadaver tissue.

ACL and PCL reconstruction surgeries have a significant rehabilitation period afterwards. However, not all patients elect to have reconstruction surgery performed. If a patient is not actively participating in sports, they may find that the occasional instability with stairs or unlevel surfaces is something they can live with and many may decide that reconstruction surgery is not for them. Young patients or athletes, on the other hand, often opt for reconstruction surgery with an orthopaedic surgeon specializing in sports medicine.

Key Points For This Chapter:

- **Osteoarthritis is the wear & tear type of arthritis**

- **Rheumatoid arthritis is caused by the body's immune system attacking the joints**

- **Other forms of arthritis include post-traumatic arthritis, infectious (septic) arthritis, and late changes from other diseases (avascular necrosis)**

- **Osteochondral defects and/or loose bodies can result in loose fragments within the knee joint**

- **Chondromalacia refers to softening of the articular cartilage the coats the joint surfaces, and this can range from mild to severe arthritic changes**

- **Meniscal tears occur when the cartilage gasket between the bones of the knee joint rip or fray. These come in a wide range of shapes and sizes, with symptoms ranging from mild to severe mechanical symptoms such as locking or buckling.**

- **Avascular necrosis (osteonecrosis) results when a segment of the bone near the joint begins to die. In severe cases it can result in collapse of the overlying joint surface.**

- **Tendinitis and bursitis are caused by inflammation of soft tissues about the knee and often resolve with conservative treatment.**

- **Ligamentous injuries to the knee include collateral ligament, ACL, and PCL sprains and ruptures. Some are treated nonoperatively and others are treated with surgery.**

Chapter 15 - Diagnosing Knee Disease

Most knee diseases are diagnosed with a thorough history, a straightforward physical examination, and routine x-rays (radiographs). Blood tests and additional imaging tests are often not required for most diagnoses. However, MRI is particularly helpful for examining the soft tissues (e.g., ligaments and menisci) in the knee, and for this reason MRI is more commonly used for diagnosing knee problems than it is for hip problems.

History and Physical Examination

As with most joint problems, an orthopaedic surgeon will typically ask questions about the involved joint, activity levels, and symptoms. In most offices, an initial intake questionnaire usually covers most of the basic questions and medical history. The surgeon usually looks this over first, and x-rays may be ordered before or after examining the patient. Nearly all patients having hip or knee surgery will have an x-ray beforehand. X-rays (also known as radiographs) are also usually obtained before an MRI is ordered.

The history is often the most informative part of the interview, and most surgeons have a fairly good idea of the diagnosis (or a short list of possible diagnoses) based on the history before even examining

the knee or obtaining any x-rays. Surgeons usually ask about the location, severity, and frequency of the pain, along with what sorts of things bring it on and what makes it better. Specific questions with knee problems may deal with previous history, particularly if there has been any prior surgery or accidents, or with narrowing down specific complaints. It is important to know if the knee is experiencing mechanical problems in addition to the pain, such as buckling, catching, or locking. A history of redness or rash may suggest an infectious problem, such as Lyme disease. All of these details are important.

The physical examination usually focuses on the affected joints themselves and adjacent joints, checking range of motion and function. Neurologic and vascular function are usually noted. Peripheral vascular disease can cause leg pain (known as claudication) and also can present problems for surgery. There are many provocative tests and maneuvers used during a physical examination to further narrow down the particular source of the problem. Pain along the joint line is often indicative of arthritis and meniscal tears.

Do not be surprised if your surgeon watches how you walk in and out of the office. As with hip problems, gait abnormalities are often very suggestive of the problem.

It is important to note that many patients who present with knee pain may actually have referred pain from the hip. Although the knee may be where the pain seems to be focused, these patients often have stiffness around the hip (particularly with rotating the hip joint, as with putting on shoes or socks) and a limp that is suggestive of hip problems. I see patients every month who arrive reporting both severe knee pain and that they have been told that their knee x-rays are fine; it sometimes turns out that these patients actually have severe hip arthritis when we get an x-ray, and they usually find that the knee pain goes away after treatment of the hip problem.

Radiographs (X-rays)

Plain x-rays of the knees are usually taken to evaluate for arthritis, fractures, congenital anomalies (such as excessive bowing or a shallow groove for the patella), tumors or metastatic disease, and other conditions. There are many things that the surgeon will be evaluating, often focusing on the appearance of the joint itself. The cartilage that coats the surfaces of the joints is transparent on the x-ray, but if the gap is not apparent, "bone-on-bone" arthritis can be seen. The gap typically is in the range of an eighth to a quarter inch, representing a cartilage layer between the bones, but it may be narrowed or even obliterated on one or both sides of the knee with severe arthritis. Other features of arthritic joints include subchondral sclerosis (hardening of the underlying bone), osteophytes (spurs), loose bodies, and cysts in the bone.

For most knee problems, it is important that weightbearing knee films be obtained. These frequently need to be repeated if a patient arrives with nonweightbearing knee films and arthritis is a potential diagnosis. There are also specific views of the knee

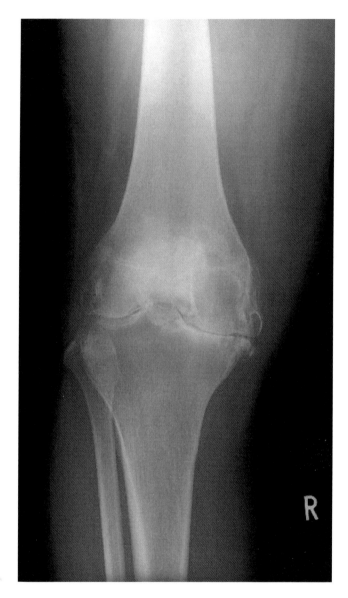

Figure 15-1. X-ray showing severe degenerative joint disease. The cartilage space between is completely obliterated with a "bone on bone" appearance.

(such as flexed 30 degrees or a "sunrise" view of the patella taken with the knee flexed to evaluate the undersurface and tilt of the patella).

Magnetic Resonance Imaging (MRI)

MRI may sometimes be ordered to evaluate for soft tissue problems (such as muscle injury, ligament injury, meniscal tear, evaluation of soft tissue mass, etc.) or for bone marrow problems. It often will "light up" for increased water content, signaling edema and injury. Bone bruises and stress fractures show up in this manner. Avascular necrosis (osteonecrosis) is often diagnosed on MRI much earlier than when it appears evident on regular x-rays.

A "torn cartilage" usually refers to a meniscal tear. The three dimensional cross sections show most (although not all) meniscal tears. These come in a variety of shapes and patterns. Anterior cruciate ligament tears and other ligament injuries also are often confirmed with an MRI.

Although MRI is often very good at imaging problems within the knee, it is not perfect. Sometimes an MRI arthrogram is required to see it well. This involves injecting the knee joint with a contrast dye in order to see tears more clearly. This is most often used if there has been a prior surgery within the knee.

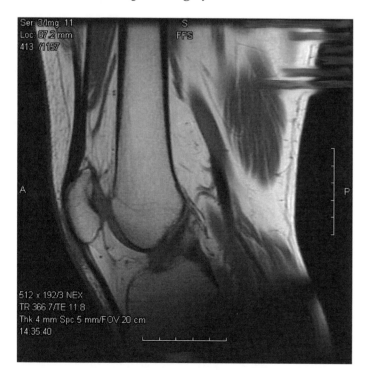

Figure 15-2. An MRI showing a cross-section of the knee from the side. The patella is on the left, and the muscles and ACL are seen in cross-section.

Computed Tomography (CT)

A computed tomography (CT) scan uses many x-ray "slices" to examine cross sections of a body or limb. The patient lays on a table while moving through a ring that contains a spinning x-ray camera. While it does have applications in trauma and spine surgery (especially when examining complex pelvic fractures), CT scans are not typically used to evaluate for arthritis. CT scans are most often used in the knee when evaluating complex fractures and trauma, such as a tibial plateau fracture (a severe fracture involving the joint surface of the tibia). Multiple small fragments and complex fracture patterns with a shattered knee can be seen on the many cross-sectional images and 3D reconstructions, aiding the surgeon in preoperative planning for surgical repair of these traumas.

Nuclear Bone (Technecium) Scans

Bone scans involve administering a very small amount of radioactive material via IV, then using a camera to view how it is taken up and eliminated by the tissues. Areas with high uptake, such as a tumor, infection, or fracture, will often "light up." This test is also useful for determining if an old hip or knee replacement is loosened from the bone, although it will provide a false positive if a bone scan is obtained within about 1 year or less of the surgery.

White Blood Cell (WBC) Scans

A tagged white blood cell scan is a similar test to a bone scan, except white blood cells are taken from the body and "tagged" with a tiny amount of radioactive material. It is then re-injected, and the scanner shows where all of those tagged white blood cells congregate in order to localize an infection. This is used when trying to find an infection in the bone or around an artificial joint.

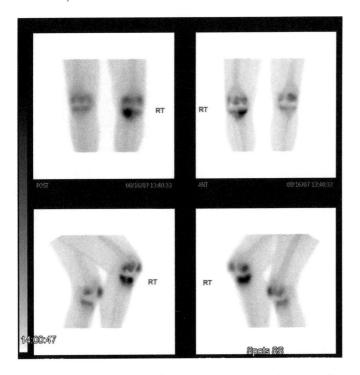

Figure 15-3. A nuclear bone scan showing increased uptake (from arthritis) around the knees.

Ultrasound

An ultrasound examination uses sonic waves to make a picture of soft tissues. While it may be used to examine cysts or other soft tissue abnormalities (such as checking to see if an Achilles tendon is ruptured), its most common application in orthopaedics is to detect the presence of blood clots in deep veins. It is sometimes used to check the soft tissues around the knee for cysts, tumors, fluid collections, or aneurysms.

Knee Aspiration (Arthrocentesis)

Sometimes it is useful to draw joint fluid out of the joint with a needle for laboratory analysis. This is most common in situations where there has been or there is suspicion for an infection in the joint itself. Usually the skin is anesthetized with a local anesthetic and the needle is placed into the joint. Unlike hip aspiration, this typically does not require x-ray to perform. The aspirated fluid is sent to the laboratory for analysis to see if there is any evidence of bacteria or infection. Knee fluid is also often checked for crystals, which can be indicative of crystalline diseases such as gout or pseudogout (calcium pyrophosphate crystals).

Key Points For This Chapter:

- Most knee problems are diagnosed through the history and physical examination

- Radiographs (x-rays) are usually all that is needed to image the joint for most conditions

- MRI is used for examining soft tissues (like meniscal tears or ligament injuries) or bone marrow (for detecting avascular necrosis or occult – hard to see – fractures)

- CT Scans are used to examine bone in detail, especially when 3D images are needed as in trauma cases such as a tibial plateau fracture that shatters the surface of the joint

- Bone scans are used to detect loosening of old implants, stress fractures, or infections in the bone

- Ultrasound is used to examine tendons, rule out blood clots in deep veins, or to see cysts or aneurysms in the back of the knee

- A knee aspiration is sometimes used to draw fluid out of the knee joint for testing, such as for infection or gout.

Chapter 16 - Nonoperative Treatment of Knee Arthritis

Conservative treatment is initially indicated for nearly all patients with knee arthritis, just as it is for hip arthritis patients, with surgery being reserved for those patients in whom conservative measures are no longer enough. Most patients have usually progressed through the spectrum of conservative treatment by the time they are referred to an orthopaedic surgeon, but it is important to be aware of the nonoperative options that can often suffice for months or years before pain and disability are significant enough for knee surgery.

Note that most of this chapter discusses knee arthritis. Some other knee problems, such as internal derangement with frequent painful locking, may need arthroscopy or more urgent treatment. That is discussed separately in the arthroscopy chapter.

Activity Modification

Many patients with mild arthritis of the knees cope initially by making simple adjustments to their usual activities, avoiding the things that aggravate the knee. It is important to maintain as much activity and joint motion as possible, but impact activities will aggravate arthritis. Running and jumping will often accelerate cartilage loss from the joint. Using an elevator instead of stairs and avoiding uneven terrain are helpful. However, there is significant evidence that remaining active and keeping the knee moving will prolong its life.

Many patients worry that they should give up walking or other low impact activities in order to try to preserve the knees, but a sedentary lifestyle actually will shorten the life of the knees, just as it will for the hips. The key is to focus on low impact activities, such as swimming or cycling. These are the best forms of exercise with arthritic knees as they do not require significant weight bearing across the hip joints. For patients who do not have access to a pool or a stationary bicycle, leisurely walking will also maintain knee range of motion, strength, and function.

Canes

Canes or walking sticks are useful, particularly when the arthritis affects only one side. Some canes have multiple feet or prongs (e.g., a quad cane) to increase stability for patients with poor balance. It is important to use the cane in the opposite hand from the bad hip or knee. This allows you to lean away from the bad leg, taking weight off of it. Adjust the height of the cane so that the hand height rests comfortably along your side, preventing stooping or poor posture.

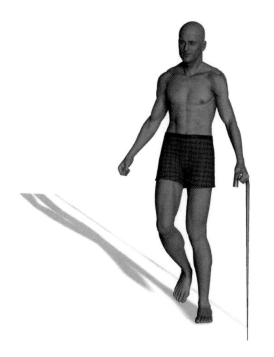

Figure 16-1. A cane can be helpful for getting around with a bad knee. Use the cane in the opposite hand from the arthritic (or recently operated) side.

Weight Loss

As discussed in the chapter on nonoperative treatment of hip arthritis, significant weight loss for obese patients can make a dramatic difference, although in actuality, relatively few patients are successful in losing weight because arthritis limits their ability to exercise. Weight loss is probably the single most effective intervention the patient can undertake on their own. Increasingly, severely overweight patients (300+ lbs.) are turning to bariatric surgery (e.g., gastric bypass surgery) with promising results, although it remains a serious operation. For the average patient who is somewhat overweight, losing 20 lbs. or more can often at least improve their discomfort.

Weight loss is also important for increasing the life span of a joint replacement. Although in our practice we do regularly perform joint replacements for patients even over 400 lbs., it is with the thorough understanding that their joint replacements may wear out more quickly, and they are at increased risk for complications with surgery. Surgery usually takes longer and is more challenging for the surgeon when the patient is morbidly obese (body mass index > 35), due to the loss of anatomical landmarks, prolonged exposure and closure time, and need for additional assistants at the time of surgery. This is true for both hip and knee replacement surgeries.

Knee Injections - Steroids

Steroid injections into the knee joint and/or bursa (between the muscle layers on the outside of the knee) are usually quite effective, and may be all that is required in combination with physical therapy and anti-inflammatory medications for resolution of a patient's symptoms. These injections do not require live x-ray, and the injection is usually administered over the side of the knee or from the front (along the joint line).

Occasionally, some patients may develop large effusions around the knee joint (e.g., "water on the knee"). This large fluid collection is an abnormally large collection of normal joint fluid in most cases, although effusions can also be seen with gout, pseudogout, trauma or injury, or infection. Patients usually feel much better when the fluid is aspirated (drained), and the fluid can also be sent to the laboratory for further analysis if needed. Many physicians will also inject the knee with corticosteroids at the same time after aspirating (as long as there is no suspicion for infection).

Injection therapies do not "cure" the underlying problems of arthritis, but can be useful for short term relief (potentially for a few months) and for diagnostic purposes. Injection of the knee joint typically is quick and only takes a few minutes in the office, and unlike hip joint injections, it does not require the use of a live x-ray machine (fluoroscopy) to ensure that the injection is placed into the joint itself. Most surgeons inject a local anesthetic and steroid mixture (contrast is not usually used, but it is employed for hip injections). The anesthetic will often make the knee immediately better and somewhat numb for a few hours afterwards, and then the local anesthetic wears off. The steroid component may take 5 to 7 days to fully take effect.

While injection of the arthritic knee joint itself is not a cure, it does have several very useful roles. It is very useful for diagnostic purposes to help determine where a patient's primary source of pain is originating from. Frequently patients may present with both knee and hip arthritis; prior to planning joint replacement, it is useful to see if their pain improves (even for just a short while) by injecting the knee. Another common scenario is that of a patient who knows that he or she needs a knee replacement but is looking for a few months of temporary relief (e.g., they are traveling, or have a daughter's wedding coming up, etc.). Most surgeons try not to use steroid too frequently as it does have some side effects (notably, weakening of the bones and tissues, and rarely, infection), but commonly surgeons will consider injections a few times per year to be acceptable.

Knee Injections - Hyaluronate

Steroid injections in the knee joint have been used for decades. A newer option is hyaluronate injection, which involves injecting a thick, clear gel into the knee joint that acts as a cushioning lubricant and anti-inflammatory agent. The injections have many different trade names, depending on the manufacturer, but most are similar in mechanism and overall effect.

Commonly used hyaluronate injections include Synvisc™, Supartz™, Hyalgan™, and Euflexxa™ injections (all slightly different preparations from different manufacturers).

Hyaluronate, or hyaluronic acid, is a substance that occurs naturally in the joint. It is a viscous substance that normally lubricates the joint. Injections of large quantities of the material, obtained either from chickens or as a bioengineered product of bacteria (similar to the way that insulin is now manufactured), often decreases inflammation and makes patients feel better. In our practice, we administer hundreds of the injections each year, and generally about two-thirds of patients report that they feel better after the injections.

Benefits of the injections include a low risk of side-effects. Repeated steroid injections can lead to serious side effects if they are administered too often, including osteoporosis. In contrast, hyaluronate injections have not been shown to have such effects, although a small percentage of patients may experience swelling. The most common downsides appear to be simply noticing no significant benefit (for patients with severe degenerative disease), cost (the injections are far more expensive than steroid injections), and the fact that most commercially available preparations require a series of weekly injections rather than a single injection. Typical series include weekly injections for 3 to 5 weeks with significant improvement usually taking about a month, and the injections can be repeated at 6 month intervals under most insurance plans and Medicare.

Given the cost of the injections, some insurance plans will not allow coverage for the expensive hyaluronate injections unless a patient has documented use of multiple anti-inflammatory medications, physical therapy, and previous steroid injection(s) that did not provide adequate relief.

Nonsteroidal Anti-inflammatory Drugs (NSAIDs)

This family of medications includes aspirin, ibuprofen, naprosyn, and other non-narcotic medications to decrease inflammation. They remain the mainstay of preoperative management of arthritis pain and are usually most useful in the early years of developing arthritic pain for all joints.

Most patients experiment with different over-the-counter NSAIDs before finding the one that seems to work best for them. Older NSAIDs such as aspirin and ibuprofen have been around for many years, and newer drugs in this class called COX II inhibitors, such as celecoxib (Celebrex), valdecoxib (Vioxx – now discontinued), and meloxicam (Mobic) have recently been introduced. Many physicians feel that these are not much different from aspirin and ibuprofen in effectiveness, although these medications have fewer gastrointestinal side effects such as ulcers. For this reason these more expensive drugs are usually employed when a patient cannot tolerate traditional NSAIDs like ibuprofen, typically because of GI upset. Some of these drugs were in the news a few years ago (notably, Vioxx) because there was some concern about heart problems in a small number of patients. These drugs also require monitoring of liver function if taken for a long period of time.

It is important not to take NSAIDs on an empty stomach, or to use them with blood thinners (such as warfarin) unless directed by a physician. Collectively, these medications are responsible for many cases of GI bleeding and ulcers in elderly patients each year. These medications can interfere with kidney function and may lead to swelling in the legs. These medications can also interfere with some blood pressure medications, and it is important to also check with the physician prescribing the blood pressure medication before taking any of these medications.

Although orthopaedic surgeons may provide an initial prescription for a month or two of NSAIDs, it is usually preferable to obtain these from your family physician over the long term because of the need for monitoring after several months of use. Some of these drugs require monitoring and periodic blood tests after prolonged use.

Glucosamine / Chondroitin Sulfate

Glucosamine chondroitin is a "nutraceutical," essentially a supplement that is often found in the vitamin aisle of the drug store or supermarket. As such, it does not typically have to adhere to the same labeling rules as drugs that are regulated by the FDA, and it is not uncommon to see labels proclaiming that it will "re-grow cartilage!" There is not much evidence that it is likely to do anything so dramatic, although there is compelling evidence that it is relatively safe and works by decreasing inflammation in the joint, making at least some patients feel better. Patients with a shellfish allergy should use caution when taking this, as it may cause an allergic reaction. The typical dosage is about 1500 mg of glucosamine and 1200 mg of chondroitin sulfate daily. Most manufacturers sell the two mixed together in a single pill. It is not uncommon to have to take it for two weeks or more before a significant benefit is seen.

Narcotics ("Pain Killers")

Most hip and knee surgeons feel strongly that these do not have a role in the preoperative management of arthritis, and in our practice, we typically do not prescribe them except after surgery or fracture. Narcotics (such as oxycodone, hydrocodone, oxycontin, etc.) are useful for treating significant pain that is expected to get better in a few weeks. When taken for a long period of time, they can have serious side effects, including addiction, constipation, confusion, and a need for higher levels of narcotics to maintain the same level of pain relief. Additionally, patients who have been on narcotics for any significant time prior to surgery are typically much more difficult to keep comfortable after surgery because they have developed a tolerance to opiates (narcotics).

Key Points For This Chapter:

- Activities that aggravate knee arthritis should be avoided if possible.

- A walking stick or cane used in the OPPOSITE hand can be helpful.

- Weight loss is important both before and after knee replacement; it may significantly prolong the life of the joints.

- Injections into the knee joint (steroid or hyaluronate) or bursa (steroid) may provide months of temporary relief, but steroid should not be administered too frequently.

- Anti-inflammatory medications like aspirin or ibuprofen are helpful, but must be used responsibly to avoid GI problems, bleeding, kidney or liver problems, or interfere with blood pressure medications. For these reasons, it is best to check with your family doctor if you take them for a long period.

- Glucosamine Chondroitin appears to help with joint inflammation and pain, but usually needs to be taken for 2 weeks or more before a benefit is seen.

- Narcotics are best reserved for surgery or broken bones (i.e., problems that are acute and will improve in a few weeks), not for chronic problems like arthritis.

Chapter 17 - Knee Arthroscopy

At times there is no substitute for simply looking inside of a knee joint to find and correct a problem. Knee arthroscopy is a minimally invasive technique for doing just that; it refers to a broad category of surgical procedures (typically outpatient) that involve looking within the knee joint using a fiberoptic video camera and small instruments. Knee arthroscopy is useful for the diagnosis and treatment for a number of common problems within the knee.

Indications

Many patients may have internal derangement of the knee, which refers to the knee having a problem from within that keeps it from moving and functioning in the normal manner that it should. Common symptoms include locking, catching, clicking, popping, and buckling or giving way. These types of problems are often caused by a problem within the joint, such as a loose body or a meniscal tear (commonly referred to as a "torn cartilage"). Arthroscopy is useful not only for diagnosing these problems by looking inside the knee, but also for correcting them by removing a loose piece of bone, trimming or repairing a torn area of cartilage, or removing excessive synovial lining or scar tissue.

There is a limited role for arthroscopy with arthritis. In the past, some surgeons have advocated "cleaning out" an arthritic knee (e.g., arthroscopic lavage), and in fact arthroscopy is very good at relieving mechanical symptoms such as locking even in an arthritic knee. Flushing the knee does remove inflammation, at least temporarily. .However, it cannot "cure" arthritis, and for that reason many surgeons will not recommend its use for simple arthritic changes unless there are also mechanical symptoms.

Arthroscopy also can be used to wash out an infection within the knee joint, without requiring a large incision and opening the entire joint.

Chondroplasty refers to smoothing out roughened joint surfaces or cartilage, and sometimes isolated "craters" in the joint surface can be drilled during arthroscopy to try to get them to fill in with scar tissue. These procedures are fairly quick outpatient procedures, but they may require limited weightbearing for a few weeks using crutches while the cartilage heals. Arthroscopy also may be used to visually inspect a knee to help determine the extent of arthritis and damage, particularly for helping a surgeon to determine if a patient may be best helped by a partial versus total knee replacement.

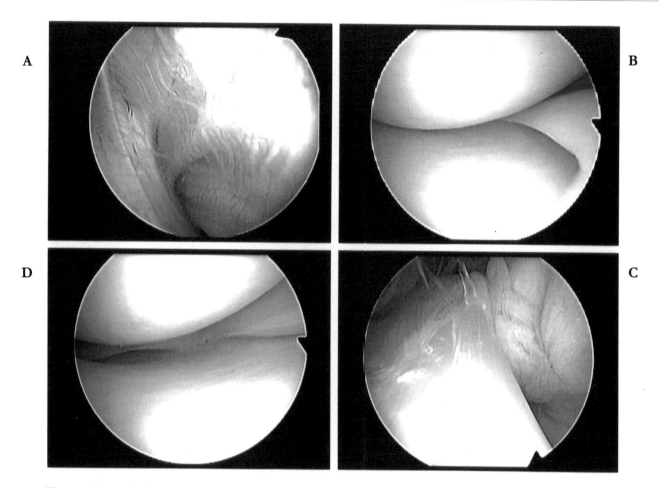

Figure 17-1. Arthroscopic inspection of healthy structures in the knee. From top left going clockwise, these show (A) the gutters and synovial lining of the joint along the sides of the knee, (B) a normal medial compartment with intact medial meniscus and healthy joint surfaces, (C) an intact anterior cruciate ligament (ACL), and (D) a normal posterior horn of the medial meniscus (femur on top, tibia on bottom, and meniscus between).

Occasionally there may be a discrete area of severe damage in the form of an osteochondral defect (see previous chapters). An isolated lesion can sometimes be treated with microfracture or arthroscopic drilling, techniques that are designed to stimulate formation of scar tissue to fill in the defect. This typically requires a period of limited weightbearing while it heals.

Meniscal tears and loose bodies are frequently seen on imaging studies such as an MRI. However, not all meniscal tears or loose bodies can be detected with an MRI, and the "gold standard" for diagnosis is to then perform a diagnostic arthroscopy. The idea of looking inside an area of the body with a camera is not limited

to just orthopaedics; similar procedures include endoscopy (looking down the esophagus with a fiberoptic camera) or colonoscopy (from the other end!), and like those procedures arthroscopy is typically a short, outpatient procedure that often does not require full anesthesia and can even be performed under local anesthesia only in many cases.

Some meniscal tears are amenable to repair, but many have a "shredded" appearance or may not be good candidates for repair because of their location (some regions do not have good healing potential because of the way the blood supply is arranged). In these cases, the rough edge is resected back to a

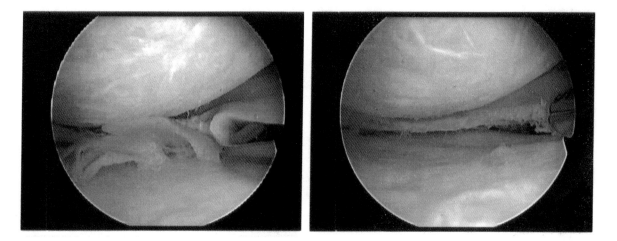

Figure 17-2. Arthroscopic partial meniscectomy. The torn meniscus on the left was causing significant pain and locking of the knee. In the photo on the right, the same tear has been resected back to a smooth, stable edge.

smooth, stable edge, which usually makes a significant improvement in symptoms. In past decades before fiberoptic arthroscopic instruments were available, surgeons would often remove the entire meniscus through an open procedure, but this often led to severe arthritis over the following years. Arthroscopy allows surgeons to remove just the torn portion of the meniscus, leaving as much as possible to cushion the joint.

Technique

The surgeon uses portals to insert the camera and/or instruments rather than opening the entire joint (which is called an arthrotomy). These portals are typically less than a quarter inch in size, and multiple small incisions may be used to access the entire joint. There are usually at least 2 portals, one for the camera and the other for whatever instrument the surgeon needs (such as a hook probe or a grabber to retrieve a loose fragment of bone, a mechanical shaver to trim torn cartilage, etc.).

Although full general anesthesia is sometimes used, particularly for complex arthroscopic procedures that are expected to take a while, most arthroscopic procedures are performed using a combination of local anesthetic injection and light sedation. It is similar to that used when setting fractures or having wisdom teeth extracted. Occasionally, some patients are interested and want to watch their surgery on a video screen as it is performed, and this is possible using local anesthetics for noncomplex procedures. Most simple procedures, such as trimming a torn meniscus or removing a loose body, take about half an hour or so.

During the procedure, water (saline or a mixture of salts and lactate called lactated ringers) is pumped through the camera and into the joint, then out through the same instrument or an additional outflow portal. This allows the surgeon to wash away any blood or material removed (such as resected meniscal trimmings or loose bodies). The fluid also fills the joint space so that there is plenty of space to safely work with the small instruments. Although an effort is usually made to drain the knee afterwards, some fluid remains within the joint. As a result, there may be minor drainage from the portal sites (usually closed with a single suture each or sometimes just an adhesive dressing) for a day or two, and this is not a reason for concern.

Most modern arthroscopy instruments have the capability to take photographs (shown here) or record live video of the image that is displayed on the video screen. Many surgeons will keep these photos for later reference in case the patient has any problems with the knee in the future, such as arthritic pain, and the surgeon can tell with a glance at the pictures what areas of the joint were noted to have the most arthritis or other changes.

When the procedure is complete, most patients spend a short while in the recovery room until they are fully awake and comfortable. The vast majority of arthroscopy patients go home the same day, and for this reason arthroscopy in many places may also be performed at an outpatient ambulatory surgery center as an alternative to the traditional setting of the hospital.

Risks and Complications

Arthroscopy is a minimally invasive and outpatient procedure. As a result, complications are rare, but there are some problems that can sometimes occur that we usually warn our patients about. Anesthesia problems are uncommon but are of particular concern for patients with other medical problems (such as sleep apnea or lung disease). Infection is rare and occurs in less than 1% of patients undergoing arthroscopy, but it can be a potentially serious complication, requiring additional surgery to treat a deep infection. Infection is somewhat more of a concern for patients who have problems with their immune systems, are on immunosuppressive drugs or prednisone, or are

diabetic, but even in these cases it is uncommon. Many surgeons give prophylactic antibiotics in the operating room just before the procedure to decrease this risk even further.

Blood clots, or deep venous thrombosis, are a rare complication after arthroscopy but have been reported. For that reason, many surgeons may recommend simply taking enteric coated aspirin for a couple of weeks after surgery. However, the current guidelines from the Academy and other authorities put the risk of blood clots at such a low percentage that prophylaxis is not strongly recommended on a routine basis.

Some surgeons use a tourniquet during the procedure. Because most arthroscopic procedures are fairly short, it is uncommon to have any problems from having a tight tourniquet placed for too long, although it has been reported. Similarly, although all patients have some degree of swelling after the surgery, in rare cases there can be too much swelling from having some of the saline extravasate, or travel into the soft tissues around the knee, and cause excessive swelling. When this occurs, it usually is immediately noted before leaving the operating room, and usually resolves within hours. Rarely, a problem from severe swelling called a compartment syndrome can result.

Injuries to nerves or ligaments are rare. When this does occur, it usually is a result of stretching the knee in order to get the camera and instruments into a hard to reach spot (usually the back of the knee). This can sometimes cause numbness or tingling, and rarely a footdrop (inability to flex the foot upward at the ankle). When these problems do occur, it usually is from a neuropraxia or stretching of the nerves, since they are not directly at risk for being cut, and this most often resolves in days or weeks. Also, a skin nerve called the infrapatellar branch of the saphenous nerve, which supplies skin sensation over the upper shin, is sometimes cut while placing the small incisions for the portals. It does not control any muscle function, and when this injury occurs the small area of numbness over the upper shin is usually not problematic.

Recovery And Outcomes

As noted, the vast majority of arthroscopic procedures are performed on an outpatient basis and are fairly minor procedures that take about half an hour or so to complete. Recovery and outcome depends on the exact problem being treated, but for the majority of procedures patients can expect to return to work within one to three weeks (depending on the procedure and what they do for work). Exceptions to this include more extensive arthroscopic procedures and those which require a prolonged period of limited weightbearing on crutches (such as arthroscopic drilling of a crater in the joint surface, which needs some time to allow the crater to fill in with scar tissue).

Key Points For This Chapter:

● Arthroscopic surgery involves looking inside the joint with a small camera .

● Most arthroscopic procedures are short and performed on an outpatient basis.

● Arthroscopic surgery can be used for diagnosis and treatment of a variety of problems within the knee, including meniscal tears, loose bodies, determining the extent of injury or arthritis, or washing out inflammation or infection.

● Arthroscopy is often performed under light sedation or even local anesthesia alone for minor procedures.

● Complications are rare from this minor, outpatient procedure but can include anesthesia problems, infection, blood clots, excessive swelling, or nerve injury.

● Depending on the exact arthroscopic procedure performed, most are quick outpatient procedures and patients are back at work within 1 to 3 weeks. Some procedures may require limited weightbearing with crutches afterwards, as directed by your surgeon.

Chapter 18 - Knee Replacement (Arthroplasty)

Total knee replacement has not been in widespread use quite as long as hip replacement, but it has a long track record and is now quite common. The first designs that resembled the modern knee replacements used today appeared in the 1970's and rapidly went through a number of evolutionary changes. It has now reached the status of a mature technology in the past couple of decades, and it is widely accepted in the orthopaedic literature that most patients undergoing total knee replacement can have an expectation of at least 95% success rate at 10 year follow-up or longer.

The basic concept of a total knee replacement (also known as total knee arthroplasty) is to replace the rough, irregular surfaces of the ends of the bones (the femur and tibia) with new surfaces. This eliminates the "bone on bone" changes from severe arthritis and allows the ends to glide smoothly over one another, with artificial surfaces that have no nerves in them. The undersurface of the patella (knee cap) may or may not be replaced also with a plastic button.

These new surfaces resemble a metallic cap that is affixed to the ends of the bone (most often with cement, although press fit cementless prostheses are sometimes used). For this reason, although "total knee replacement" has been the term used for several decades, "knee resurfacing" would probably be a more accurate description since it is usually half an inch or less that is actually removed from the ends of the bones and replaced. It is not unlike a dental procedure in which a bad tooth is capped. After the joint is replaced, there is no longer any arthritis in the joint, because the joint surface is entirely artificial.

Partial knee replacements also exist, most often as a unicondylar knee replacement, which replaces one side of the knee only. These are less invasive procedures and typically have a quicker recovery, with the advantage of retaining more "factory original" parts. However, only some patients are candidates for a partial knee replacement. It will only help the portion of the knee it replaces in most cases, and if both sides of the knee joint are worn out, it is often better to consider a total knee replacement. Some patients also have significant deformity or angulation, making it difficult or impossible to correct alignment and biomechanics without a total knee replacement.

At the time of surgery, the ends of the thigh bone (femur) and upper leg (tibia) are typically quite worn out. Frequently, the ends of the joint look very similar to two heads of cauliflower in a very worn out knee, covered with lumpy and bumpy osteophytes (spurs) and areas of exposed bone, grinding against each other.

Nuts & Bolts: Total Knee Replacement Procedure

Regardless of the surgical approach used, the same general steps have to be performed during the surgery. Some surgeons use a tourniquet for the procedure, and others prefer to identify transected blood vessels and ligate them at the time of surgery (rather than have them bleed into the joint after surgery when the tourniquet is released). Tourniquets can also be a source of soreness and circulation problems after surgery, and for that reason we typically do not use them in routine knee replacement surgeries in our practice.

After exposing the knee joint - usually with a vertical incision in the front of the knee - the irregular, arthritic ends of the femur and tibia are resected. These cuts are made in a way to keep the mechanical axis of the knee properly aligned, which usually requires keeping the perpendicular cut at about 5 to 7 degrees off of the vertical axis. Because the end of the femur is rounded (i.e., shaped like a cam mechanism), it is also necessary to make chamfer cuts. These are usually made in such a way that the new "cap" fits very tightly over the chamfer cuts. Remnants of the menisci and anterior cruciate ligament, if they are still present, are removed.

Next the upper end of the tibia is resected. It is important for the surgeon to cut and prepare this surface at the proper angles also; if the cut is tilted too much side to side, the knee will either be excessively bowed or knock-kneed. Similarly, if it is angled too far up or down when viewed from the side, knee flexion and extension may be adversely affected. It is also important for patients with severely bowed legs to understand that full correction of the deformity may not be possible at the time of surgery.

The surgeon checks that the knee is "ligamentously balanced" at this point, which is the most difficult part of the biomechanics to restore. For example, the knee might be too tight when flexed but loose and unstable when fully extended, or vice-versa. Although there are a wide variety of knee replacement designs used to address different problems, the majority of knee replacements are designed to keep most of the patient's own ligaments, which keep the knee stable when moving back and forth and from side to side.

Many combinations of biomechanical challenges need to be resolved at this point to make sure the knee moves in as natural a way as possible, which is one reason why knee replacements are arguably more technically demanding than hip replacements.

A polyethylene (special plastic) spacer is also selected to fit between the two metal components in most knee designs. This has a very low friction surface that allows the new knee replacement parts to smoothly glide over one another. Some knee replacement designs utilize an "all-polyethylene" tibial component, which is all plastic without a metal backing.

These spacers have many different geometries and sizes, depending on the biomechanical needs identified by the surgeon. Some are designed to replace the function of the posterior cruciate ligament, and others are designed to work with an intact posterior cruciate ligament. Some are designed to allow more range of motion and others favor more stability. As with most things, there typically are some trade-offs made in order to find the best replacement for each patient. Trial components are usually used before cementing in the final components, which allows the surgeon to check the range of motion, knee tracking, and ligamentous stability before implantation of the final (real) components.

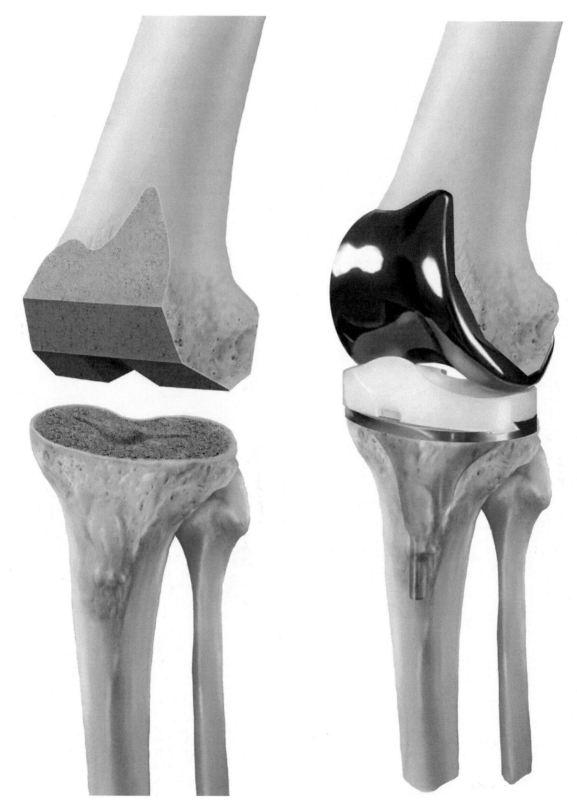

Figure 18-1. The arthritic surfaces of the bone are cut away (left) and the new knee replacement components fit securely over the ends of the bone (right). Note that the knee replacement on the right is black and specular; this is an Oxinium(TM) implant made from oxidized zirconium.

(Courtesy of Smith & Nephew, reprinted with permission.)

The undersurface of the patella (knee cap) may or may not be resurfaced. There are some surgeons who always resurface the patella, and some who never resurface it. Many surgeons decide at the time of surgery whether it is warranted. Replacing the undersurface of the patella with a small plastic button replaces one more arthritic surface, but it does have the potential to introduce mechanical problems with tracking of the knee cap and also is another cement interface that will eventually wear out over the years. For this reason, many surgeons will not replace it unless they feel it is warranted.

The artificial parts are typically cemented into place with polymethylmethacrylate bone cement. While there has been a definite trend in *hip* replacements to utilize cementless components that are porous and allow bone growth into the prosthesis, most surgeons have departed from cementless *knee* replacements. Early results in the 1990's showed a higher incidence of early failure and loosening in the knee when cement was not used, and this has been thought to be the result of the the different biomechanics and forces seen in the knee as opposed to the hip. However, some researchers have begun advocating cementless knee replacements again with newer data, and this trend may reappear in the future.

Antibiotics are sometimes mixed into the cement if there is concern about the patient having an increased vulnerability to infection. These antibiotics typically are gradually released out of the cement into the joint and offer a protective effect for several months.

After the new knee is solidly fixed in place and tested with the final parts cemented into place, the knee joint is then typically closed in multiple layers with various sutures. There are several ways to close the skin, ranging from staples to traditional sutures to absorbable sutures with special surgical glue. In our practice, we tend to use absorbable sutures with glue for most patients since we have found this heals quite nicely and does not require removal of any sutures or staples. A drain may or may not be placed depending on surgeon preference and the degree of bleeding noted during the case. Some surgeons may also inject the knee at the end of the case with various pain medications. We typically inject the knee with morphine and a long-acting anesthetic (marcaine) as we finish the closure.

Unicondylar Knee Arthroplasty (Partial Knee Replacement)

A unicondylar knee arthroplasty refers to replacing just one side (usually the inner or medial side) of the knee. It has been somewhat controversial since it was introduced several decades ago, with many surgeons praising it as a less invasive alternative to total knee replacement, and others critical of early results that showed relatively short longevity. However, research and efforts through the 1980's and early 1990's began to show results and implant survivorship closer to that of total knee replacements, and interest was renewed.

Today, most joint replacement specialists find that partial knee replacement is an attractive option for the right patient. It is a less invasive procedure that preserves the ligaments and soft tissues of the knee in addition to only replacing a small amount of bone in the area where it is most degenerated. Patients typically rehabilitate significantly faster than with a total knee replacement, and we often see shorter hospital stays for partial knee replacements (typically 48 to 72 hours for most patients). However, most joint replacement specialists would likely agree that the key is choosing the right patient for the procedure.

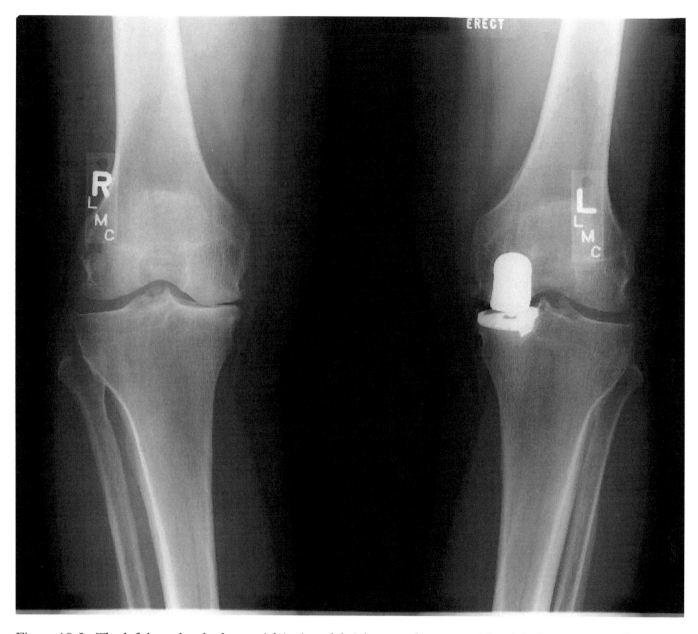

Figure 18-2. The left knee has had a partial (unicondylar) knee replacement. The right knee is nearly "bone on bone" along the inner aspect, and will likely need the same surgery soon.

Clearly, partial knee replacement is not as good as a total knee replacement if more than a single compartment of the knee is severely worn out. It works best for the patient who has significant degenerative changes and pain on just one side of the knee, usually the inner (or medial) side. In my own practice, I will usually recommend a total knee replacement if a patient has pain and joint line tenderness on *both* sides of the knee.

Other considerations include the patient's size and degree of deformity. Although it is possible to correct some degree of bowing with partial knee replacement, severe angular deformity or instability often necessitate reconstruction with a total knee replacement. It also tends to be less optimal for very obese patients, who may benefit from the increased durability of a total knee replacement.

Conversion of a partial knee replacement to a total knee replacement is relatively simple, and the procedure does not take significantly longer than a first-time (primary) total knee replacement. The most common reasons in our practice for eventual conversion surgery are either progression of arthritis over the years in the parts of the knee that were not yet replaced or loosening of the prosthesis. These are important factors to be aware of, and patients who receive partial knee replacements need to know that eventual conversion at some point may be necessary. However, for a younger patient, this leaves more of the bone available at the time of revision, and for that reason is often particularly attractive for a younger patient (under 65) who will probably need multiple eventual revision surgeries in their lifetime.

Patellofemoral Knee Arthroplasty (Partial Knee Replacement)

Another form of partial knee replacement is to replace only the undersurface of the patella and the groove that it rides in along the front of the femur. The replacement part in the groove is a smooth piece of metal that is inlaid flush with the surrounding bone/cartilage. Patellofemoral knee arthroplasty is not often used, as it is uncommon for a patient to have a problem with only the knee cap and its groove that is severe enough to warrant surgery without also having worn out the rest of the joint (usually to the point of needing a total knee replacement). However, it may have a role in very select cases.

Key Points For This Chapter:

- In total knee replacement (total knee arthroplasty), the ends of the femur and tibia are replaced with artificial parts. The undersurface of the patella may or may not be replaced.

- In a partial knee replacement (unicondylar knee arthroplasty), only one side of the knee is replaced. This typically allows faster recovery and retention of more of the patient's own knee, but patient selection is critical. It is not a good procedure for the patient who has worn out multiple compartments of the knee, has severe deformity or angulation, or is severely overweight.

- Patellofemoral arthroplasty is not used very often, but involves replacing only the undersurface of the patella and its groove.

Chapter 19 - Surgical Approaches For Total Knee Replacement

Knee replacements are carried out through one of several different surgical approaches. In practice, however, there is not as much difference between knee surgical approaches as there is between hip surgery approaches. There are advantages and disadvantages to each approach, and there is some controversy among knee surgeons as to which is the best. All surgeons have a favored surgical approach, and while there are often heated and spirited debates at academic conferences and meetings, it is a testament to the success of the procedure that all of them generally produce good results.

Recently there has been much interest in minimally invasive surgery for knee replacement, as there has been with hip replacement surgery. There is no widespread agreement as to exactly what constitutes "minimally invasive," but most surgeons would agree that the general principal is to have less soft tissue disruption and dissection. As with the hip, this often may translate into a smaller skin incision, but it is what goes on under the skin that is far more important for speed of recovery. The single most important difference for minimally invasive knee surgery appears to be decreasing the amount of muscle and tendon that is disrupted.

Anterior Surgical Approach

Most of the surgical approaches for the knee have a similar skin incision, going vertically over the front of the knee. The reason for this is primarily that most of the important blood vessels and nerves are in the back of the knee and are avoided. It is also important to have an *extensile approach*, meaning that the incision can be extended upwards or downwards as needed.

The traditional anterior surgical approach goes straight down the middle of the knee. The tight capsule around the knee, called the retinaculum, is usually opened along the inner (or medial) side of the patella (knee cap). This is often called a "medial peripatellar arthrotomy," meaning an opening into the knee joint is made just along the inside of the patella. The patella is then moved out of the way to expose the knee joint. Traditionally, this involved everting (flipping) the knee cap, but recent studies have shown that patients rehabilitate faster if the surgeon takes care to simply slide the knee cap to the side. It does not offer as much visualization, but we find in our practice it is worth doing it this way in order to promote a faster recovery, and most surgeons who perform many knee replacements a year have a good understanding of the anatomy that facilitates such less invasive techniques.

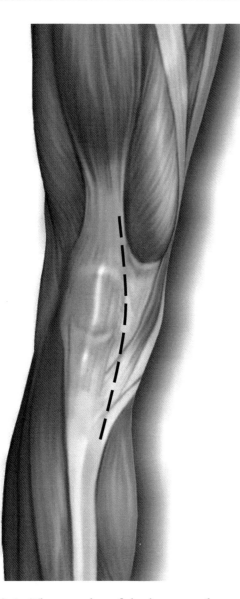

Figure 19-1. The muscles of the knee are shown, and the medial peripatellar arthrotomy most commonly used is marked by the dotted line. Note how it goes just along the patella and quadriceps.

Subvastus Approach

On the surface, this incision also typically goes vertically down the front and center of the knee, and from the outside skin incision it is difficult to tell a difference between the various approaches described here. However, the subvastus approach, first described

in 1929, employs a slightly different approach once under the skin layer. Instead of making an incision along the patella, this approach lifts the vastus muscle up and over the knee. It requires starting far along the inner aspect (medial side) of the knee and elevating the entire front muscle mass up and over to expose the joint. Advantages include less muscle disruption and possibly better patella tracking by leaving the extensor mechanism intact. However, disadvantages include less visualization and access to the joint, and because of the nature of the approach it is not well-suited for patients who are not thin. Access to the joint can be problematic if the patient is at all obese, has a tight knee or contractures, has significant bowing or deformity, or has had any previous surgery. For these reasons, it is not used as commonly as the anterior/medial peripatellar approach described above.

Midvastus Approach

The midvastus approach is very similar to the anterior approach described above, except that in the muscle layer (the skin incision again is the same) the incision turns away from the center of the knee and avoids cutting into the quadriceps tendon. This approach gained popularity in the 1990's because it offers some of the advantages of both the anterior and subvastus approaches, with good visualization and access but better preservation of the extensor mechanism. Advocates of this approach believe that avoiding cutting into the quadriceps tendon leads to a rapid restoration of post-operative extensor mechanism function and knee range of motion. This is the technique most often used in our own practice, and we are advocates of its use in evolving minimally invasive techniques.

Previous Incisions and Scars

Previous incisions and significant scars have a significant impact on the surgeon's decision of exactly how to make the incision and approach. The primary reason for this is the blood supply to the skin and soft tissues around the knee. Although an old surgical scar may be long since healed, that previous scar has disrupted the blood supply. As a result, if a parallel incision is made too closely, there may be significant wound healing problems because of a diminished blood supply. In general, surgeons will typically either try to incorporate an old incision if possible, or make a new incision at a right angle to it to minimize the effect of the old scar.

Another consideration is scar tissue that may make the knee very tight. The surgeon needs to be able to get within the knee joint to work on the bone surfaces and put in a new knee replacement (and possibly remove an old one in revision surgeries). If the knee is too tight to allow this, then several options exist. The surgeon may disconnect the area of bone below the knee cap where the tendon attaches (called the tibial tubercle) and re-attach it at the end of the case, or alternatively the quadriceps tendon may be cut at the upper end of the knee and re-attached afterwards. While these techniques make the surgery possible, they do add significantly to the recovery time and often require that the knee be immobilized in a brace for weeks or months while the repair heals.

Commentary

As noted at the introduction to this chapter, there is not as much variability in surgical approaches to the knee as there are to the hip. In fact, even among different surgical approaches, the outside skin incision typically will look the same (e.g., a vertical line down the front of the knee). The differences arise among the way that the underlying muscles are split.

As greater interest has evolved in minimally invasive surgery, a major factor has been changes in the instrumentation actually used during the surgery (retractors, insertion devices, jigs, etc.) that allow significantly shorter incisions today (and more importantly, less disruption to underlying muscle and tendons). Other changes with less invasive techniques have included not everting the patella throughout the case and minimizing damage to the quadriceps tendon.

Advances in knee surgery in recent years have mostly been evolutionary rather than revolutionary, given that the original designs over 30 years ago were functionally quite good, and refinements have been gradual. Materials science and engineering have advanced, and the knee replacement prostheses used today have a typical lifespan measured in decades.

Computer Navigation?

Industry and implant companies have developed a number of different technologies in the past 10 years that allow the surgeon to use a computer system during surgery that is coupled with some sort of positioning system. Commonly, a set of video cameras are positioned around the operating table as the computer system's "eyes," and it monitors the position of instruments and the patient's bones by using some sort of landmarks (often markers that are fixed to the bone temporarily during surgery).

The value of computer navigation during surgery has been debated extensively in recent years among surgeons. Many surgeons and centers specializing in joint replacements have found that the added operating time for setting up the cameras, "registering" the patient's bones and anatomical landmarks, and using the software and navigation system adds significant operating time without clearly improved outcomes. There is concern that this may translate into increased complication rates from increased time on the operating table (e.g., infection, blood clots, etc.). At the time of this writing, some studies show it to be helpful while others do not.

Navigation systems may prove helpful to surgeons who do not frequently perform joint replacement surgery and assist in accurate placement of the components. However, it has been our experience and that of many other large joint centers that outcomes are not significantly improved with navigation. This technology will very likely mature in the years ahead to the point at which it is not as cumbersome, and computer navigation may become more commonplace. For now, some surgeons find it useful and others find it unnecessary.

Conclusion – Multiple Surgical Approaches Exist

In summary, there are multiple surgical approaches for knee surgery just as there are for hip surgery, and there are also multiple surgeons who advocate one particular approach over others. However, surgical approaches to the knee actually differ very little from each other, without the significant differences seen between hip surgery approaches. At our center, we do a great deal of research and publication regarding minimally invasive surgical techniques and feel that the anterior approach has a strongly proven record of good outcomes over the past three decades, and we are currently using the midvastus technique for many

minimally invasive surgeries, but there are proponents of all surgical approaches at various centers.

As with hip surgery, the best advice for the patient is to find a surgeon whom you like and feel comfortable with, be sure that he has good surgical outcomes and a significant volume of knee replacements (preferably multiple knee replacements each week, rather than several per year), and allow the surgeon to use the surgical approach and technique that he is most accustomed to.

Key Points For This Chapter:

- **There are multiple surgical approaches to the knee, and all get the job done with a high degree of success. There is not as much difference between surgical approaches to the knee as there are with approaches to the hip, and from the outside the skin incision looks very similar for all of them (vertical skin incision over the front of the knee).**

- **There is disagreement among knee surgeons as to the "best" approach, with advantages and disadvantages to each approach.**

- **We use the anterior and midvastus approaches because of their versatility, good outcomes, and the advantages of minimally invasive surgery with rapid rehabilitation.**

- **Ultimately, all surgeons should use the surgical approach that they personally get the best results with. The best predictor of patient outcome may be the volume of knee replacement surgeries the surgeon does each month.**

Chapter 20 - Surgical Alternatives To Knee Replacement

The vast majority of patients reading this book with an interest in knee surgery will be considering either total or partial knee replacement. However, there are some other surgical alternatives that exist for arthritis. Most of these options were developed in the years before joint replacement or resurfacing was widely available, but some are still performed today in select cases.

One notable exception is knee arthroscopy, which is a versatile procedure with many applications that involve looking inside the knee. For that reason, it is discussed in its own chapter.

Knee Fusion (Arthrodesis)

In the years before knee replacement, a common surgery for a severely arthritic knee was to fuse the femur to the tibia, usually with a long rod through the center of the bones, effectively eliminating the joint and creating a single bone from the thigh to the ankle.

This surgery persisted even after the development of knee replacements, primarily as an option for young patients (such as laborers) who would otherwise wear out an artificial joint very quickly. Modern designs and materials have largely made this surgery obsolete, however, and it is very rarely considered as a first

(primary) surgery for arthritis in this country. Few patients today in America would be willing to accept the limitations of a fused knee. It is problematic with sitting at a movie theater or on an airplane, and even getting in and out of a car can be difficult.

The principal disadvantage of a fusion is that there is no longer any motion at the knee, given that the bones are fused together. This leads to an awkward gait pattern. Sitting and walking are severely affected. Also, the back and the hip typically begin to develop arthritis from "double duty" trying to accommodate the lost motion.

However, once the two bones have fully grown together, a fusion will rarely need any further medical treatment. There is no implant to wear out, break, or become infected. It is also much cheaper than using a knee replacement prosthesis. For this reason, this surgery is still used in poorer parts of the world where knee replacement is not an option for patients. In this country, it is primarily used today as a salvage option after failed or infected joint replacement surgery.

Osteotomy

An osteotomy involves cutting the bone and re-aligning it to heal in a different position or angle. You could think of it as a controlled, surgical fracture. Several types of osteotomies have been used over the last century for the treatment of arthritis and other knee problems. Note that osteotomies are used for different bones in the body for treatment of a variety of problems.

One particular application that is still used for early knee arthritis is a high tibial osteotomy, which involves cutting the upper end of the tibia and re-aligning it in such a way as to take weight off of the worn out side and to increase weightbearing on the "good" side of the knee. This changes the angle of the leg. Therefore, to be considered a candidate for the surgery, most patients need to have severe arthritis in only one side of the knee with preservation of the joint on the other side. It is most often considered for a young patient with heavy physical demands, such as a heavy laborer, who traditionally has been a difficult to treat candidate for replacement surgery because of the demands placed upon the knee.

Most osteotomies take a long time (months) to heal. Studies have also shown increased incidence of complications when the osteotomy is later converted to a knee replacement, making that eventual surgery significantly more complex and riskier. Increasingly, many orthopaedic surgeons and the orthopaedic literature in general are considering partial and total knee replacements with modern materials and designs to be a better option than osteotomy, but there are still many advocates of its use, and for certain patients with specific needs it may still be an attractive option.

Interpositional Devices

When one side of the knee joint wears out and the leg becomes bowed, some surgeons have tried inserting various interpositional devices into the knee to "shim" it back to normal alignment. Essentially, this involves making a small incision and inserting a small disk-like device (typically metal) into the worn out side of the knee. This was initially tried many years ago and abandoned by most surgeons in favor of partial or total knee replacement, although periodically newer designs resurface with renewed interest.

While the various "spacer" devices often do work in correcting alignment, they typically have not been very effective with pain relief. Most orthopaedic surgeons recommend partial or total knee replacements as a more proven technology.

At the time of this writing, there is some renewed interest in this type of surgery with custom-made inserts. Several companies are offering smooth metal spacers that are custom made based on 3D CT scan or MRI data for individual patients. Once the custom made spacer is manufactured, it is inserted in a very short (possibly outpatient) procedure through a small incision and using an arthroscope. While it is an attractive concept, there is not much experience or data on these new devices yet.

Knee Arthroscopy

Knee arthroscopy involves looking inside the knee with a small camera. Frequently in this type of surgery instruments are used via small incisions, typically less than a quarter of an inch, to remove debris or loose bodies, trim torn cartilage, smooth rough cartilage, and perform a variety of other tasks without a large incision. An entire chapter in this book is devoted to this category of surgical procedures.

While there has been widespread use of arthroscopy to "clean out" arthritic changes from the knee, many joint replacement specialists find that it is best used for accomplishing specific tasks, such as finding a loose body that is causing locking or buckling. While arthroscopy is a great tool for meniscal tears, loose bodies, and other problems, it has a limited role for arthritis treatment.

I explain to my patients that arthroscopy does not "cure" arthritis, and at best it may temporarily improve their discomfort and symptoms for a while, but the best use for outpatient arthroscopy with an arthritic knee is to fix a mechanical problem such as locking. It can also be used to smooth out some isolated areas of roughened cartilage (e.g., chondroplasty), but it does not cure or reverse generalized arthritis. It also can often be used to inspect and photograph the inside of a knee joint to decide whether a patient is a good candidate for a partial versus total knee replacement.

Key Points For This Chapter:

- **Knee fusion (arthrodesis) is the process of fusing the femur to the tibia, eliminating the painful joint but also eliminating motion. It is not used much today in this country except as a salvage procedure for failed joint replacements, although it is an alternative to knee replacement that is still used in poorer countries.**

- **Knee osteotomies (cutting and re-aligning the bone of the tibia, femur, or both) are a potential alternative that predated joint replacement, especially for young laborers with severe arthritis affecting only one side of the knee, but require a long time to heal and can lead to a higher incidence of complications.**

- **Interpositional devices have been tried with limited success and are not widely utilized at this time, although there is some renewed interest with custom made devices machined from CT scans or MRIs for individual patients.**

- **Knee arthroscopy uses small incisions to insert cameras and instruments into the knee to inspect and photograph the joint, remove loose debris, smooth roughened cartilage or trim torn cartilage, and other tasks. It does not "cure" generalized knee arthritis, although many patients will see a temporary benefit.**

Chapter 21 - Knee Prosthesis Designs

Knee replacement surgery has not been in widespread use quite as long as total hip replacement surgery, but it is a close second, with knee replacement designs continually evolving over the past 3 decades. One reason that hip replacements predated knee replacements is the increased complexity of the knee joint; the human knee is a significantly more complicated structure than the hip, and the biomechanics are correspondingly complex.. The earliest designs treated the knee as a hinge (and some salvage-type prostheses used for tumors or other large surgeries still work in that manner), but it turns out the knee is actually more complicated than that, resembling a four-bar linkage with some additional complexities such as the screw-home mechanism and femoral roll-back.

As with hip replacements, there are numerous different designs and variations on knee replacement prostheses. This chapter discusses the most common features of the principal types of prostheses. Materials are also discussed.

Decision To Select The Implant

There are many factors that the surgeon must consider when choosing an implant. Firstly, there are many different options today, including cemented versus noncemented components, cruciate retaining and posterior stabilized designs, mobile bearing knees, partial and total replacements, and several different choices for bearing surfaces. Secondly, there are advantages and disadvantages to each of these choices, and the rest of this chapter presents the most common pros and cons of each design choice. Thirdly, it is a fact of life that cost is increasingly becoming a factor in this aspect of the surgery, and in many places around the country, surgeons are at least somewhat limited by hospitals (and in some regions, limited quite severely) to what choices of implants can be considered. In some cases, the newest and best prostheses may cost more than the hospital will be reimbursed by insurance or Medicare, and the hospital may have policies in place to limit the use of such implants.

In general, operating rooms may often allow orthopaedic surgeons who perform a large volume of joint replacement surgeries more latitude because of the numbers of patients in which the hospital does not take financial losses. Surgeons performing a small number of joint replacements each year may not have that flexibility at the community hospital. We have been fortunate in our institutions to have the ability to choose from a wide variety of implants for our patients.

Partial (Unicondylar and Patellofemoral) and Total Knee Replacements

As discussed in previous chapters, some patients are candidates for partial knee replacements, while others need a total knee replacement. The most common types of partial knee replacements replace the inside (medial) side of both the tibia and femur surfaces. These types of replacements, called unicondylar knee replacements, typically have metal surfaces (usually titanium or cobalt chrome) that are fitted to the tibia and femur surfaces with cement, and some sort of polyethylene (plastic) bushing fits between them to decrease friction. Less commonly, a patellofemoral replacement may replace only the undersurface of the knee cap (patella) and the groove that it travels in.

A total knee replacement is so named because it completely replaces the contact surfaces over the ends of the femur and tibia. This also typically has some sort of low friction polyethylene bushing placed in between. There is considerable variety among total knee replacement designs.

Cemented And Noncemented Knees

In the early days, all knee replacements were secured into the bone with cement. This is still the case for most knee replacements, for mechanical reasons that are discussed in the knee arthroplasty section. The cement itself has changed very little, and polymethylmethacrylate (PMMA) cement is utilized in all sorts of orthopaedic applications where cement may be needed. In actuality, it functions more as a grout than true cement, filling in the porous spaces around a prosthesis.

The cement offers the advantage of initial strength. At the time the patient leaves the operating room, it is as strong as it will ever be. This is also the downside, given that cemented components eventually loosen like a cobblestone in a cemented walkway. Although noncemented prostheses are now outperforming many cemented prostheses in hip replacement surgeries, this has not proven to be the case with knee replacements. Noncemented knee replacements are "press-fit" over the ends of the bones, and the bone grows into the porous surfaces on the backsides of these devices. While somewhat more popular in the 1990's, noncemented knee replacement designs are not currently favored by most surgeons because of early loosening that was seen in these devices, and good results have been consistently seen with cemented designs (often lasting 15 to 25 years).

Another advantage of cement is the the ability to mix antibiotics into it, like vancomycin or gentamicin. These antibiotics are released by the cement over a period of months. Sometimes antibiotic-impregnated cement beads or spacers are placed temporarily into a joint in order to fight infection for weeks or months.

Introduction to Bearing Surfaces

The ends of the femur and tibia are replaced with artificial components, much like capping a tooth in dentistry. Most of the time the endpieces themselves are made from metal, typically cobalt chrome or titanium (and in the case of the tibial component, occasionally completely from polyethylene).

The bearing surface, however, can be made from a different material than the endpiece components. This allows the surgeon to have some flexibility in using the best material for the bone interface (such as titanium, which is structurally strong but does not work well as a bearing) and the best material for the bearing itself (such as cobalt chrome or zirconium oxide). There are a number of different materials now used for the bearings, but the most common are metal on plastic (a cobalt chrome femoral surface on a plastic spacer, which can be exchanged years later if it wears out), hybrid material on plastic (such as using a zirconium oxide ceramic surface, such as Oxinium™, that moves on a plastic spacer), or using an all-polyethylene (plastic) tibial component.

Cruciate Retaining vs. Posterior Stabilized

Most patients are surprised to learn that the majority of the ligaments in the knee are retained after knee replacement surgery. While there are some designs (such as constrained designs used in revision surgeries) that do assume the function of the body's ligaments, for most primary (first-time) knee replacements it is advantageous to keep many of the patient's ligaments to make the knee function more anatomically and to keep it stable.

Nearly all knee replacement designs do assume that the anterior cruciate ligament (ACL) is no longer viable, as it often is torn or absent by the time knee replacement surgery is needed, but most also assume that the medial and lateral collateral ligaments (MCL and LCL, the ligaments that keep the knee from moving side to side) are still intact and retained.

A principal difference is whether or not the knee replacement keeps or replaces the posterior cruciate ligament (PCL). This ligament is normally responsible for keeping the knee from sliding backwards under the thigh, particularly when going up or down stairs and inclines. A cruciate retaining design assumes that the PCL is left in place, and it is thought to offer some advantages in reproducing the knee's natural motion (particularly femoral rollback). A posterior stabilized design uses a plastic post in the center of the knee to replace the PCL, which may be needed if the posterior cruciate ligament is loose, torn, or absent.

In some cases, it may be desirable to replace the posterior cruciate ligament if the patient has an inflammatory arthritis (such as rheumatoid arthritis) in which the PCL may be expected to deteriorate over time. However, a posterior stabilized design does have the disadvantages of requiring more bone removal from the femur and uncommonly resulting in the knee "jumping the post," in which the femur may jump over the post (or cam) and essentially dislocate, requiring a visit to the hospital to relocate the knee.

Some surgeons routinely use one type of design or the other, and although there are advantages and disadvantages, both types in general have very good outcomes and few mechanical problems. In our practice, we now often use the posterior stabilized design, particularly in patients who have an absent PCL or have an underlying disease process that may predispose to ruptures of the PCL in the future.

Mobile Bearing Knees

Mobile bearing knees have been used for a number of years, but are not used in the majority of knee replacements. These designs employ a slightly different design philosophy, allowing the plastic (polyethylene) bearing surface to spin. The theoretical advantage of this is to separate the motions for flexion and extension from rotation, possibly improving bearing wear and biomechanics.

Over time, however, these advantages have not been clearly identified in most studies, and the majority of orthopaedic surgeons performing joint replacements find there is little difference in the long term performance of these designs from standard designs. However, mobile bearing knees can on infrequent occasions "spin out," similar to a dislocation. For this reason, we typically do not employ them in our own practice, although it should be noted that surgeons in some centers have reported good results with these types of designs.

Constrained Knees, Augmentations, And Stems

Revision knee replacement surgeries are significantly more complex than first time knee replacements. Besides requiring removal of scar tissue, previous components, and any remaining cement, these procedures often have special requirements, such as dealing with knee instability.

Revision knee replacement designs may include a number of features not seen in most primary (first time) replacement prostheses, including "constrained" knee designs. These designs make up for a lack of knee stability by having matching curved surfaces that do not allow much give outside of normal flexion and extension. In cases of severe instability, such as patients facing revision surgery after multiple previous surgeries or large excisions (such as with tumor surgery), a hinged prosthesis may actually be used. These do not recreate the motion of the knee as well as those designs used for primary surgeries, but they can be used to treat difficult problems with knee instability.

Some patients have large cysts present (seen at the time of surgery or on their x-rays) that may be large enough to plan on grafting and filling at the time of surgery. Some bones have cysts (hollow areas) or thinned areas of weakened bone that will need to be examined and possibly grafted. If areas of bone are significantly weakened, then struts of graft bone may need to be wired or cabled around the shaft for additional support.

Sometimes the underlying bone may not be as strong as it needs to be, or in some cases, large defects (e.g., holes) may exist from previous surgeries, infections, or trauma. These can be filled in with metal augmentations, such as metal wedges that fit under the joint surfaces to fill and "shim" the missing area. Sometimes long stems may be needed, similar to the keel under a sailboat, that anchor the knee replacements into the femur and/or tibia. These are commonly used in revision surgeries.

Other Factors – Geometry and Gender Specific Knee Replacements

This chapter is by no means an exhaustive list of the different factors used in implant selection, but it acquaints you with the major design differences and the general benefits and drawbacks of each.

The surgeon also considers a number of other factors in implant selection as well. The geometry of the femur varies considerably from patient to patient; some patients have a narrow shape to the femur and others are broad. There are prostheses that fit each of these and other geometries.

Some orthopaedic implant companies have recently introduced "gender specific" knees to try to address geometry differences seen between men and women. This topic is quite controversial in contemporary orthopaedic surgery at the time of this writing; while most surgeons acknowledge that there are general variations in the shapes of the bones between men and women, many surgeons view these new labels (and corresponding large national advertising campaigns) as a marketing "gimmick" to try to drive patients (consumers) to request the use of various manufacturers' implants. As a result, many patients are learning about these offerings through television and magazine advertisements as companies try to increase their market share. Some companies are marketing "knee replacements for women," and not surprisingly, women make up the largest percentage of patients undergoing knee replacement surgery.

Many surgeons, including myself, view the "new" concept of "gender specific" knee replacements with some curiosity. Even between members of the same sex, there are tremendous variations in the size and shape of the implants needed; for this reason, it is more accurate to say that knee replacements need to be "patient specific." In fact, this is the case with most of the modern knee systems, in which large differences in size and shape can be accommodated by selecting the right implant for the right patient. Men may have a small femur and women may have a broad femur, and vice-versa. It is more important to get it right for the individual, by taking the time to carefully measure and template x-rays before surgery and to make careful measurements and implant selection during surgery, rather than to rely on a shape and size that is estimated to work better for one gender over the other.

In the end, there is a long list of factors that can play a role in the implant selection, but your surgeon is trained to consider the pros and cons of each and uses these factors to reach a decision for which implant to use. An educated patient is the best patient, but consider that there is not that much information conveyed in a 30 second television commercial urging you to ask your surgeon to use a particular type of implant. However, your surgeon has spent years studying the different types of implants. If you have questions about why he believes a particular implant is best for you, then you should certainly ask your surgeon.

Key Points For This Chapter:

- Partial and total knee replacements exist.

- Cemented prostheses use bone cement to fix the prosthesis to the bone. Noncemented prostheses are porous coated so that the bone grows into the prosthesis. Most knee replacements used today employ cement.

- Metal on plastic (polyethylene) bearings are the most commonly used. Alternatives include hybrid materials such as a zirconium oxide surface on polyethylene.

- Many different total knee replacement designs exist, including cruciate retaining, posterior stabilized, and mobile bearing knee replacements.

- Revision knee replacements may use constrained shapes, metal augmentations, bone grafts, stems that are anchored into the femur and/or tibia, and even hinges.

- Gender specific knee replacements are controversial at this time. A more accurate statement would probably be that knee replacement needs to be "patient specific," picking the right size, shape, and type for the individual patient. If you have questions, ask your surgeon about the implant selection.

Chapter 22 - Hospitalization For Knee Replacement

Typical hospitalization times usually range from two to four days for most total knee replacement surgeries, with the partial (unicondylar) knee replacement patients usually needing slightly less time in the hospital by a day or so. The events leading up to and surrounding the surgery are similar for all joint replacement surgeries and are detailed in a later chapter, including information about preoperative testing, blood donation, anticoagulation, etc.

In general, most of the patients treated in our practice can expect a general timeline similar to the one presented here, although there certainly can be variability depending on other medical issues and the exact type of surgery performed.

Day of Surgery

Most patients arrive at the hospital early on the morning of total and partial knee replacement surgeries. This process is detailed in a later chapter because the process is the same for both hip and knee replacement surgeries. The surgery usually takes about 60 to 90 minutes for routine primary (first time) surgeries, although it certainly can take longer if the patient has had prior surgery, is very muscular or obese, is having more than one procedure performed, etc. In our

practice, partial knee replacements tend to take a little less operative time than a total knee replacement.

The time in the operating room for most surgeries is actually longer than the time needed for the procedure itself, because the anesthesiologist needs time to perform the spinal or general anesthesia, the patient needs to be positioned and prepped with antiseptic scrub, and sterile drapes have to be placed.

Most patients are in the recovery room for two or three hours after the surgery, at which time routine x-rays are taken to check on the implants and surrounding bones. Patients also need to be fully awake and have a stable blood pressure before being transferred to their hospital room upstairs. However, many knee replacement surgery patients can actually get up with assistance that first evening, just to stand or take a few steps.

After Surgery

Physical therapy starts in earnest on the second day of hospitalization. Most patients are allowed to fully bear weight on the affected knee (or knees, if both were replaced at the same time), but it is important to work with the physical therapist to ensure that there is no dizziness from medications or anesthesia. We typically allow most patients to fully weightbear right away if the postoperative x-rays show everything to be in good position and there are no special circumstances. Most surgeons (as in our practice) prefer to start full weightbearing as soon as possible, as this has been demonstrated to have a positive effect on recovery. Your surgeon and physical therapist will let you know your weightbearing status after surgery.

Most patients use a walker at first, and when they are ready and steady enough, they progress to a cane. There is a great deal of variability in how long the process takes, because everyone has different levels of physical stamina, balance, muscle strength, etc. As a general rule, younger and more active patients and thinner patients will graduate to using a cane quicker than their older or heavier counterparts. Patients who have had surgery on just one leg will progress more quickly than patients who have had both knees operated on.

Most IV lines, urinary catheters (if you have one – not all patients require this), and surgical drains (again, if you have one) are removed on the first or second day after surgery. Most of our patients do not have any drains, but they are occasionally employed depending on the size of the patient and surgical factors. Although many patients may be apprehensive about removal of drains or catheters, most are somewhat surprised to find that this is not usually as uncomfortable as they expected and literally takes only seconds.

There are typically daily blood tests to monitor hemoglobin levels and metabolic parameters, and patients are monitored to make sure that they do not have any complications after surgery such as a blood clot or a serious hematoma (more on this later). Blood thinners of some sort (either aspirin, heparin / enoxaprin injections, or warfarin) are usually started the

day after surgery also (the exact regimen depends on the patient, surgeon, and medical factors). In our practice, enoxaprin injections are usually used for several weeks after significant knee surgery to prevent blood clots.

In contrast to total hip replacement patients, who have a list of precautions to follow so that the hip does not dislocate in the initial weeks while the soft tissues are healing, knee replacement patients are limited more by their stamina and ability to move on their own. Nearly all knee replacement patients are allowed to place full weight on the knee right away (unless there are special circumstances, such as a fracture), and knee replacements do not have the risks of dislocation that a hip replacement presents. One of the primary concerns with knee replacements is stiffness in the weeks and months after surgery, and for this reason range of motion exercises are extremely important. The physical therapist works with you to practice your recommended exercise regimen and learn what you should and should not do after leaving the hospital.

Some surgeons and hospitals employ an automated machine called a continuous passive motion (CPM) machine, which is a mechanized cradle that fits under the leg and slowly bends the knee back and forth while the patient is in bed. There are pros and cons to using such a device, and in general, most surgeons would agree that working with a physical therapist or nurse is a better work out for the knee. Additionally, we would prefer that the knee muscles are actively moving if possible. CPM machines are frequently used when a patient is either unable to get up and fully participate because of other medical issues or when there is a less than optimal amount of physical therapy time and therapists available.

Leaving The Hospital

After two to three days, most partial knee replacement patients are typically ready to graduate to home with visiting nurses and therapists. Discharge to a rehabilitation facility is not as common for partial knee replacements as it is for total knee and hip replacement patients, although sometimes there are physical or social factors that make a short stay at a rehabilitation center a good idea. Common reasons may include having both knees operated on, very large patient size, patients who live by themselves, or other related factors.

After three or four days, most total knee replacement patients are ready to graduate either to home with visiting nurses/physical therapy or to a short term rehabilitation facility. This depends on how well the patient is able to get around on their own and their age, but most often the need for a rehabilitation center is dependent on social factors. If a patient lives alone or does not have adequate help, or if the living arrangements cannot be changed to accommodate staying on one level, then a short stay at a rehabilitation center is more likely.

Post-operative Visits

The first post-operative visit is usually between 3 and 6 weeks after surgery. If there are staples or non-absorbable sutures, these are usually removed by the visiting nurse at 2 weeks after the surgery. Patients are discharged with a detailed list of instructions on what to do, what not to do, and what to call about (such as fevers, wound breakdown, calf swelling, etc.). These are described in greater detail in a later chapter.

Key Points For This Chapter:

- Most primary (first time) knee replacements require 2 to 4 days in the hospital.

- Most first time surgeries take between 60 and 90 minutes per side, depending on patient size and other factors. Partial knee replacements take somewhat less operating time and these patients are typically discharged sooner.

- Patients usually arrive the day of surgery unless they have a medical problem that necessitates earlier admission (such as blood thinners that cannot be discontinued).

- Patients are usually up and walking within 24 hours of the surgery.

- Each day, mobilizing and getting up are the most important factors. It gets progressively easier.

- Most patients go home with home visiting nurses/physical therapy, but some go to rehabilitation facilities depending on age, general physical condition, and social factors.

Chapter 23 - Home Life And Exercises After Knee Replacement

It is very important to continue with physical therapy and exercises after any joint replacement procedure. Once patients are out of the hospital, the surgery may be finished but the physical therapy and rehabilitation are just beginning.

This may sound daunting to someone who is contemplating surgery and currently dealing with a painful knee, but in actuality, most patients find that the pain they have after surgery is quite different from the preoperative pain. Patients commonly remark immediately after surgery that they no longer feel the grinding, deep joint pain with weightbearing that they had previously, and that most of the discomfort after surgery is muscular pain in the area of the incision. Perhaps even more importantly, this type of discomfort steadily resolves and actually improves the more patients are up and about using their new knee.

There is little disagreement that there is definitely some pain associated with the postoperative period. Pain medication, early mobilization, and time all help to reduce this period until you are moving around well and getting out of the house on the new knee. Knee replacements do tend to be somewhat tougher in recovery than hip replacements, but over 95% of patients can expect a good outcome after 9 to 12 months after the surgery.

Although many patients are driving and getting around within a month after knee replacement surgery, I often advise my patients that they should expect about a year before the knee reaches maximum strength and range of motion.

Getting Home And Transportation

Most patients are able to go home in a regular car, and within a few days, they can certainly use the car as a passenger to get to the hospital or physician's offices. Generally it works best to use a vehicle that allows you to stretch your legs out in front of you, although this is not as important as it is for hip replacement patients (who may need to avoid flexing beyond 90 degrees).

Generally it is best to avoid nonessential travel out of the house for about 7 to 10 days after a total knee replacement. Most patients are able to go for short rides or to a restaurant after about a week. Partial knee replacement patients can expect to mobilize somewhat quicker. Although some younger and more active patients have actually returned to office (desk) jobs for short periods after the first week, it is typically best to do the exercises / physical therapy and otherwise rest in the first few days after surgery.

You should not take any extended car trips for 5 weeks. This is primarily because of the prolonged sitting and the increased risk of blood clots, the same as seen for hip replacement patients.

Driving is usually not recommended until 2 or 3 weeks after discharge, if you have good control of your right leg, and if you do not have any other medical conditions that prevent you from driving. If you have other conditions besides your knee that may impede your driving (such as low blood pressure, vision, or neurologic issues), check with your family physician before driving. Obviously, if you have lightheadedness or are still taking narcotic medications, then you should not drive.

We typically recommend that patients practice driving in an empty parking lot, such as an empty school or church parking lot on a Saturday. It is also a good idea to take a family member or friend with you, and if you BOTH feel comfortable with your ability to drive, then begin driving short distances and gradually work up to longer trips.

It is important to understand that you have to take legal responsibility for determining when you are safe to drive. If you feel you are unsafe, then wait until you feel more confident.

Getting Around On Your Own Two Feet

You should be walking at least 4 or 5 times per day, increasing your distance each time. **Walking and bending the knee are your most important exercises after a knee replacement.** It will increase your stamina and strength, decrease stiffness, help to

prevent blood clots and constipation, and you actually will feel much better if you are mobile.

However, when you are not walking, remember your rest periods in bed with leg elevation. These breaks are important to prevent swelling. You should spend most of the first week resting in bed, elevating the legs for as long as there is significant swelling. Keep the legs elevated above the level of the heart. However, while in bed, you certainly can (and should) continue to bend the knees back and forth.

Flex the ankles up and down whenever you think about it, which promotes circulation. You may walk frequently, but in general you should spend two hours, twice a day, in bed with the legs elevated for as long as there is persistent swelling in the leg. If your leg and calf suddenly become much more swollen, warm to the touch, and painful in the calf, it can be a sign of a blood clot and you should call the surgeon's office.

During the first week at home, you should not sit in a chair for more than 3 times a day for 30 minutes each time (usually at mealtime). While sitting up is good for many things, it does increase swelling in the legs, and therefore in the first week try not to sit up in a regular chair for more than 30 minutes at a time. After the first week, this can be relaxed if there is not significant swelling or discomfort. Sitting periods can slowly be increased to a normal routine after the first week.

While sitting in a chair, try to have everything you need (e.g., water, telephone, television remote, etc.) within arm's reach so that you do not twist.

Stairs

You may begin using stairs as soon as you feel comfortable. Some patients with good stamina and muscle strength may practice stairs at the hospital before going home; others will take a few weeks to build up their strength. The most important factor is to be safe, and always use a handrail for balance as you begin using stairs again. If you feel unsteady, then you may use a sitting position to scoot up or down the stairs.

When going up stairs, lead with the **unoperated** leg, and when coming down, lead with your **operated** leg. (If both legs have been operated on, then you can use whichever leg is more comfortable.)

Figure 23-1. Go UP stairs leading with the nonoperated (or stronger) leg.

Figure 23-2. Go DOWN stairs leading with the operated (or weaker) leg.

Showers and Toilets

Different surgeons may have different guidelines, but in our practice, we generally allow showering 2 days after discharge if the wound is dry. Gently towel the area dry after showering. Do not shower or get the wound wet until 2 days after the wound has become completely dry, and do not allow it to get wet if there is still some drainage.

A shower stool is a good idea for the first 6 weeks after surgery. This can be helpful to avoid slipping and falling.

No tub baths for at least 6 weeks. This is primarily to avoid the motions involved with getting in and out of a tub, but generally it also is not a good idea to completely submerge the surgical site for a couple of weeks.

You may use a regular toilet, although a raised toilet seat is sometimes helpful by decreasing the strain on the knees associated with getting on and off the commode.

Sleeping at Night

Unlike hip replacement patients, knee replacement patients do not require a pillow between the legs and may sleep in whatever position they find most comfortable. However, it is a good idea to elevate the leg and prop a couple of pillows under the calf and knee to prevent swelling, particularly in the first week or two after surgery.

Exercises

The exercise program after joint replacement is not temporary, but continuous. It is an important part of the ongoing management of your knee replacement.

As noted above, **walking is the most important exercise.** You should walk at least 4 or 5 times a day, increasing the distance each time. It is better in general to walk for shorter periods with rests in between than to attempt a marathon session once or twice a day. Rest periods are helpful in between. The walking should be at a slow, steady pace on level ground. (I often recommend going to the mall several weeks after surgery for most patients, as it is level ground and weather is not a concern.) Walking faster will not be particularly beneficial, and if you strain the muscles by walking too quickly, it is possible to have some muscle bleeding and swelling in the first week or so. A slow and steady gait, on the other hand, is very beneficial.

Some surgeons will routinely prescribe the use of a mechanical device to slowly move the knee, called a continuous passive motion (CPM) machine. We will prescribe these on occasion in our practice, but typically only for those patients who have cannot comply with exercises on their own. Active exercises and walking are more beneficial.

The following exercises are the ones we recommend in our practice, primarily focusing on walking and bending the knee. If you are reading this and have another surgeon, be sure to check with him about your exercise instructions and routine (and any special precautions or limitations) in case it differs.

Ten sets of the following exercises should be done each day, and at least 10 repetitions of each exercise should be done during each set. The standing exercises should be done while holding on to a table or using a crutch or cane for balance.

Bend the Knee And Hip

Bend both the knee and hip in a standing position, lifting the leg up and down 10 times. Hold on to a table or walker for balance. Make an effort to completely straighten the knee. You should also keep moving the knee even while seated or reclining in bed. In this way, try to bend and straighten the knee "a thousand times" a day.

Figure 23-3. Bend the knee back and forth. This can also be done while in bed or seated. Try to completely straighten the knee and bend it back as far as possible.

Ankle Motion

Do not forget to flex both ankles up and down whenever you think about it, like stepping on the gas pedal. This helps the circulation and prevents clots. It is particularly helpful while laying in bed.

Knee Isometric (Quads) Strengthening

Lying flat, keep the legs straight and a little apart. Try to tighten the thigh (quads) muscles and also push the knee downward against the floor or bed, holding the contraction for 5 seconds. Repeat 10 times.

Knee Range of Motion / Straight Leg Raise

While laying flat, place a rolled towel or round pillow under the knee, then extend the knee so that the foot rises completely off the floor or bed. Hold it up for 5 seconds, then release. Repeat 10 times.

Figure 23-4. Isometric (Quads) strengthening. Try to tighten the muscles of the thigh while pushing the knee downwards.

Figure 23-6. Straight Leg Raises. A pillow behind the thigh can be helpful while trying to straighten and extend the knee (but not directly behind the knee).

Heel Slides

Lying flat, slide the operated foot up as far as you can while keeping the heel in contact with the floor or bed, then allow it to slowly slide back. Repeat 10 times.

Figure 23-5. Heel slides. Bring the knee up as far as you can, then allow it to slide out, straightening it.

Key Points For This Chapter:

- Keep in mind that these are the post-operative recommendations in our practice and that they apply for most but not all patients; always check with your surgeon about your specific limitations and instructions after surgery.

- Walking and bending the knee are the most important exercises after surgery.

- Take frequent breaks and rest periods during the first week. Plan to spend most of the first week in bed when not exercising.

- No extended car trips for 5 weeks.

- You can start driving in a few weeks if you feel safe to do so. Practice driving in an empty parking lot first! Do not drive while still taking narcotics.

- Shower 2 days after discharge if the wound is dry and your surgeon has instructed you to do so. However, no tub baths for at least 6 weeks.

- Go up stairs leading with the unoperated leg, come down leading with the operated leg.

- Your exercise program is not temporary, but continuous. It is an important part of the ongoing management of your joint replacement.

Chapter 24 - Life After Knee Replacement

Most patients have minimal pain by 3 months (or sooner) after knee replacement, and the majority return to the office and report that their discomfort level, activity, and quality of life are all dramatically improved. It is not unusual to have occasional muscle aches and persistent (but usually slight) swelling of the knee and extremity for several months. Depending on numerous factors, including the surgical approach, amount of surgical work needed, and particularly the state of the musculature around the knee before surgery, some persistent limp is usually expected. In some cases, a limp may be persistent for a long time after surgery (most often for a patient who has had significant muscle atrophy – or wasting – from longstanding disuse of the knee prior to surgery).

It is important to remember that while most patients are back working and getting out of the house within 6 to 8 weeks, they are usually continuing to see gains in muscle strength and range of motion for up to a year after surgery.

Returning To Work

There is a wide variation in how soon patients return to work. It primarily depends on what you do and also on your physical condition before surgery.

Obviously, a young person who is in good health aside from a bad knee will be back to work much sooner than a patient who is severely overweight and out of condition, but most everyone gets there eventually if their health is reasonable and they work at the rehabilitation. Partial knee replacement patients tend to recover significantly more quickly than total knee replacement patients, and patients who have had bilateral (e.g., both knee) procedures take a little longer.

Some patients have been back to desk jobs within several weeks. Others who have very physical jobs, such as laborers, may need to take 10 to 12 weeks until they are able to meet the demands of their job. We recommend that everyone returns to work when they can function safely and with reasonable comfort. Each patient is responsible for determining if he or she can safely perform the activities of his or her job (no one else knows what the job entails better than the person who has to perform it).

If there is some accommodation by the workplace (e.g., to allow someone who normally works a standing job to do light duty at a desk for a short while), most patients are back to at least limited work by 6 weeks.

Activities after Knee Replacement

In general, it is best to avoid impact activities after knee replacement. Although some patients do activities such as jogging or contact sports, they usually do so against medical advice. Avoid situations where repetitive impacts or sudden jolts might occur. Impacts will decrease the life of the replacement and increase the likelihood of early loosening, possibly necessitating revision surgery.

Low impact activities such as walking, golf, cycling, swimming, hiking, or ballroom dancing are good sources of activity and cardiovascular exercise, and these activities are well tolerated by joint replacements (assuming your general medical condition allows them). Skiing on gentle slopes is usually safe, although we recommend against downhill skiing that involves significant twisting and turning. Most joint replacements will last for many years with proper care and low impact activities.

Some patients enjoy yoga or pilates, and these activities are usually fine for routine exercise with some modification to accommodate the range of motion recommended by your surgeon. While hip replacement patients do have some range of motion limitations, knee replacement patients are limited by their own ability (e.g., knee replacement patients do not have to be mindful about dislocation like hip replacement patients do).

Sex After Knee Replacement

Another area that patients are often afraid or embarrassed to ask about is sex after any joint replacement. Hip replacement patients have a little more restriction (see that chapter), but knee replacement patients are again generally limited by what they can do.

Generally, sexual activity is safe for both men and women once they feel comfortable enough to resume activities. Most patients can resume sexual activity 4 to 6 weeks after routine knee replacement if they are otherwise healthy enough, often sooner with partial knee replacements.

One area of caution is that of kneeling. Total knee replacement patients in particular find that kneeling is too uncomfortable, although it generally does not result in physical damage to the knee. If you have more creative positions in mind, most things can be accommodated with some common sense and by going slowly. If you have a concern about a particular activity, it is probably best to just ask your surgeon about it rather than do something risky.

Going To The Dentist

In the past, most surgeons have typically recommended the use of prophylactic antibiotics before dental procedures. The reason has been that with dental work, there is often bacteria in the bloodstream for a short while afterwards that (at least theoretically) may lead to infection of the joint replacement. This concept is not limited to prophylaxis after joint replacement surgeries; many cardiologists also recommend antibiotics for patients with heart valve problems for the same reason.

In reality, there are bacteria in the bloodstream on many occasions (such as after brushing your teeth), and the risk of infection of an artificial joint is very low.

There are however some situations in which antibiotics are recommended. The American Dental Association (ADA) and the American Academy of Orthopaedic Surgeons (AAOS) have jointly met and issued some guidelines regarding when antibiotics should and should not be used before dental work. Because this information also applies to knee replacements, there is an appendix with these guidelines at the back of this book.

Longevity of the Implants

Most patients want to have some idea of how long they can expect their knee replacement to last. This is highly variable, however, and there are many factors that contribute to the longevity of the implants used. A good analogy is that of a new car; when two different owners drive a brand new car of identical make and model off the lot, there may be a wide difference in how many years each is able to use the car based on mileage and how they drive.

Impact activities (running, basketball, and other jumping sports) will increase the likelihood of loosening for a knee replacement over time. The repetitive impacts can lead to slow but progressive loosening of the components from the bone.

Patient weight has a significant influence over how long the replacements will last. Heavier patients place a larger load on the implants, but conversely, they often are less active (e.g., take fewer steps in a year).

The type of bearing material used factors into longevity. There is some evidence that hybrid materials (e.g., zirconium oxide on polyethylene rather than traditional cobalt chrome on polyethylene) will last longer, and these more expensive implants are typically employed for younger and more active patients. At this point, however, with most revision surgeries performed in recent years we are typically seeing the surrounding bone wear out (e.g., loosening) before the implants themselves wear out.

If a part of the knee replacement does wear out, it often may be only the polyethylene liner that is worn while the metal components (fixed to the bone) are still good. In this case, most implant designs allow for a relatively quick procedure in which just the liner is exchanged.

Factors unrelated to the implants may shorten their lifespan, such as infection or trauma. I have treated patients who fell or were involved in motor vehicle accidents who fractured the bone around the implant, necessitating wiring or revision. In general, it usually takes the same amount of energy required to break a bone to damage the replacement, however.

Younger patients wear out their replacements more quickly than older patients. For this reason, many surgeons in the 1980's and 1990's recommended waiting as long as possible before replacing joints in young patients. While we still wait until all conservative (nonoperative) treatments are exhausted, we now recognize that the implants and technology have evolved to the point that even young patients can expect years of use before requiring revision surgery, and partial and total knee replacement can now give severely impaired patients their mobility and life back.

Key Points For This Chapter:

- There is wide variation between patients regarding when they return to work (several weeks to several months) depending on how physical their job is and their overall physical condition

- Most patients are driving within 4 to 6 weeks after surgery

- Partial knee replacement patients generally tend to rehabilitate more quickly than total knee replacement patients

- Knee replacement patients should avoid repetitive impact activities and sports (such as running), although low impact activities should be encouraged (walking, swimming, cycling, golf)

- Most knee replacement patients (especially those with total knee replacements) find that kneeling is uncomfortable and generally avoid it

- Sexual activity can be resumed when patients feel up to it, usually at 4 to 6 weeks

- Dental visits and antibiotic recommendations are included in an appendix at end of this book

- Longevity of the replacement depends on many factors, including activity, patient weight, bearing materials used, and external influences (traumas, infections, etc.)

PART III - BEFORE & AFTER

JOINT REPLACEMENT

SURGERY

Chapter 25 - Office Visit Before Surgery and Initial Evaluation

The majority of patients are referred to the orthopaedic surgeon by their primary care provider, although an increasing number of educated patients today make appointments directly with our office for hip and knee problems. While some HMO's may require referral to see a specialist, many patients make the appointments on their own. Some patients are also referred by physical therapists, nurse practitioners, chiropractors, or other specialists (rheumatologists or spine surgeons).

History and Physical Exam

At the first visit with the orthopaedic surgeon, a detailed history and a physical examination will be taken. The history is in many ways the most important part, and most practices that specialize in hip and knee surgery like ours will have a two or three page history and intake form. We usually mail this form to patients before surgery or ask them to print it from our website. It has many of the questions that we need to ask already included, and additionally provides important information on other health issues, allergies, and medications.

In our office and in many orthopaedic offices, the preliminary history and examination may be performed by a physician assistant. This is a professional with at least graduate level education who can help gather information, perform a physical examination, and provide many physician extender functions.

The history is supplemented by asking a number of detailed questions, primarily to gain additional insight into the exact symptoms and limitations patients are experiencing and to help the surgeon formulate a diagnosis. This interview can sometimes be completed by telephone for patients who live far away and by having the medical history and x-rays available ahead of time. However, it does not eliminate the need for a physical examination.

The physical examination in an orthopaedic office is somewhat different from a physical examination at the primary care provider's office. Much can be determined simply from observing how a patient sits, stands, and walks. This gives the surgeon a very good idea of what may be going on with the hips and knees. The range of motion of these joints will be checked, and joints are usually tested for ligamentous stability (for example, most knee ligament injuries can be diagnosed by testing each of the ligaments physically).

The joints are examined for evidence of erythema (redness), effusion (or fluid collection), signs of new or old injuries/scars, and for internal derangement. Specific tests and maneuvers may be used to check for problems such as a meniscal tear, etc.

Radiographs (X-rays)

Radiographs (x-rays) are often obtained. Not all orthopaedic problems require x-rays (such as diagnosing bursitis or tendinitis), but most arthritic conditions – especially when surgery is being contemplated – require imaging of the joints. Surgeons assess many things on x-rays, including bone quality, joint space / articular cartilage, the presence of cysts or spurs, findings of loose bodies in the joints, new or healed fractures, and anatomic deformities (such as the degree of varus – or "bowleggedness" - and other anatomical variations or evidence of congenital problems like hip dysplasia).

If you have had previous x-rays or other studies (such as MRI's or bone scans), it is a good idea to bring the actual films with you to the office. Many patients have arrived with just a copy of a radiologist's report and do not understand why the surgeon needs to see the actual x-rays or needs to repeat the x-rays in the office when they have been done elsewhere. Having a radiologist's report that simply says "degenerative joint disease" or "osteoarthritis" does not tell me as a surgeon what the exact problem is, how severe it is, and most importantly, whether surgery may be needed to correct it. That can only be determined from actually reviewing the films themselves – a picture really is worth a thousand words.

Increasingly, many radiology centers and physician's offices are able to place all of a patient's x-rays and other images onto discs that can be easily transported (much more so than a heavy jacket full of MRI films) and viewed on any computer.

In addition, we often may repeat the x-rays if the previous films are not of sufficient quality or taken with the wrong technique to show what we need to see. A very common example of this is standing x-rays with knee problems; many primary care providers may order knee films with the patient lying down, and for diagnosing many orthopaedic conditions we need the knee films to be taken with the patient standing, in order to see how much cartilage gap remains in the knee and to determine knee alignment when weightbearing.

Other Tests

Other tests may be ordered after the first visit, depending on the exact problem being investigated. Most patients needing hip or knee replacement do not require any further imaging beyond the x-rays, but we may sometimes order an MRI to evaluate specific conditions (such as evaluating for a meniscal tear in the knee – or "torn cartilage" - which will not be evident on x-rays) or a bone scan to determine if a prosthesis is loose.

We usually will have patients return a week or two after the tests to discuss the results. Some tests may be simple – such as a Lyme disease test – but most imaging tests such as an MRI will yield complex results and frequently determine what we do next. For that reason, we usually will ask patients to come back to discuss the results and the next step unless it is a simple, confirmatory test that we can relate over the telephone.

Discussing Surgery

Many patients can be diagnosed on the first visit without the need for additional tests or imaging (such as for knee or hip arthritis), and we can discuss their options on that first day. For the vast majority of patients needing hip or knee surgery, there is no immediate emergency in scheduling hip or knee replacements or arthroscopy, and we will usually present the options, discuss the surgery, and recommend that patients go home and think about it.

When patients do need hip or knee replacement, hip resurfacing, or knee arthroscopy, we will usually discuss the surgery itself, the hospitalization, the common risks and benefits, and alternatives to surgery. Clearly, this information can fill a book like the one you are reading now, and so we usually try to present the pertinent information and may provide additional reading materials.

Sometimes a more detailed visit to discuss surgery is needed. Many insurance plans will cover this, although Medicare will not in most cases. (Medicare will not typically cover additional preoperative visits to discuss surgery once it has been recommended. Unfortunately, most patients are not aware of this, but hopefully this policy can be changed eventually if enough citizens voice their concerns to Medicare about it.)

It is best to take family or friends with you to the office visit if they will be involved in your care or decision making process. This is important for several reasons; the first is that a large amount of information may be conveyed during the visit, and two people are more likely to remember it than one! The second reason is that HIPPA (Health Information Privacy and Portability Act) enacted by Congress severely limits healthcare providers' ability to discuss patient information over the telephone with anyone other than the patient or their designated power of attorney. The third reason is that other family members (usually children of older patients) may have questions, but it is difficult to answer those questions over the telephone if the patient and their medical chart are not immediately present (e.g., the day after the visit). The best time and place to ask those questions (by far!) is at the office visit with everyone present.

Key Points For This Chapter:

- **Some insurance plans or HMO's require a referral to the orthopaedic surgeon by your primary care provider; others allow patients to schedule visits directly**

- **The initial office visit will entail a detailed history and physical**

- **If you have any previous x-rays or MRI's, obtain the actual films and take them with you to the office visit! A radiologist's report does not usually include enough information to make surgical decisions.**

- **If you have obtained prior opinions or if you are seeing the surgeon for a second opinion, it is a good idea to take previous office notes (and operative reports, if applicable) with you**

- **Other tests may be ordered, such as an MRI or bone scan. Usually another visit is scheduled to go over complex test results and discuss what the next step is.**

- **Most hip and knee surgeries are elective, and you have time to think about it.**

- **Take a friend or family members with you to the office visit! It is difficult to discuss patients' surgery or care over the telephone for legal and practical reasons.**

Chapter 26 - Deciding On Surgery

Hip and knee surgery is usually elective, meaning that it is not usually life-threatening (the exception would be for hip fractures). Most joint replacement surgeries are for patients who have had time to think about surgery carefully and are well-motivated to do well with the surgery and participate in the rehabilitation afterwards.

It is common for patients to ask *when* they should have surgery for hip replacement, hip resurfacing, or knee replacement. Clearly, some conditions are urgent, such as a hip fracture or an impending fracture from a tumor, but the majority of patients are undergoing surgery primarily for pain relief, and secondarily to improve their function. Joint replacement is typically scheduled when the patient reports that the level of discomfort and impairment are such that they are willing to accept the risks of surgery to have it fixed.

As a general rule in our practice, if a patient is a candidate for an elective surgery (often either joint replacement or resurfacing), we have a long discussion with the patient about what the surgery involves, possible risks, possible alternatives, and what we hope to gain with the surgery. We usually recommend that the patient talk it over with family members, give it careful thought, and let us know if and when he or she would like to schedule elective surgery.

Take Family & Friends With You

It is often a good idea to take family members or friends along with you, particularly if they will be involved in the decision making process with you. Although it is certainly possible to have an excellent outcome without any additional help, most patients benefit greatly from the assistance of family or friends during and in the weeks after hospitalization. Moreover, the office visit is usually packed with a great deal of information from your doctor, and studies (and experience!) show that most people are unable to absorb all of it and often forget at least some details.

As mentioned earlier, another good reason to include family members or friends in the office visit is the Health Information Privacy Portability and Accountability Act (HIPPA) enacted by the federal government a few years ago. Although the law was intended to protect patients' privacy, it has had some unforeseen consequences over the past few years that have frustrated patients and healthcare providers alike. The law is quite strict about discussing patient care and information with anyone other than the patient without permission. I frequently receive telephone calls from patients' adult children or even spouses a week or two after their visit wanting to know what was discussed with the patient and details about surgery, and we are both frustrated that I cannot discuss these things with them. Additionally, when I receive their phone call days or weeks after I have sat down with the patient and discussed their surgery and plan in detail, I may not have all of that information right in front of me to answer those questions. Therefore it is best to bring all involved family members or friends to the office visit, and everyone's questions can be answered completely.

Can you wait too long?

Many patients ask if it is possible to wait too long to have the surgery. The answer to that is not simple, but the most important response is that the patient should have the surgery when he or she is good and ready for it. This includes being both physically and psychologically fit for the surgery. Conversely, some patients putting off surgery do go on to develop significant other medical problems (such as major cardiac issues) that subsequently prevent them from having the surgery, and they may wish they had had the surgery before their health deteriorated. But even so, I advise my patients not to proceed with surgery until they are comfortable with the decision to do so.

Most patients decide to have elective joint replacement when they find that they are significantly changing or limiting their daily activities in order to accommodate the disease. For many, this is primarily because of the pain, but other common reasons I hear from patients are inability to do the activities that they love (such as playing golf, tennis, going for walks, playing with grandchildren) or loss of independence.

Arthritis does not get better. If anything, the best that can be hoped for is a slow progression. Flare-ups of the symptoms may subside, but the times between flare-ups will gradually shorten, and the severity of the

episodes will increase until it becomes a constant problem for most patients. Bones and muscles lose strength and mass (osteoporosis and muscle atrophy) when joints are severely arthritic, although significant muscle recovery is possible in the months after the joint is replaced.

Putting off surgery when arthritis is severe can have some detrimental effects as deformity increases (this is especially true for knee replacements) and muscles deteriorate, leading to a more difficult rehabilitation, or in some cases, limiting how strong they will ever be. This can limit the maximum range of motion that a patient may realize after surgery. Further destruction of the joint can make the surgeon's job somewhat more challenging as well.

Lastly, I often advise patients that if they ever find themselves needing narcotics on a daily basis to get through the day, it is probably time to think about surgical intervention. Narcotics are best used for a short term problem, such as after a fracture or right after surgery, and serious problems can result with long term use.

Second Opinions

You should always keep in mind that there is never any harm in obtaining a second opinion. Some patients obtain a third, fourth, or more opinions before deciding on a surgeon and surgical procedure they are comfortable with. Few good orthopaedic surgeons will ever disagree with getting additional opinions prior to an elective procedure. However, you should probably provide your surgeon with records and office notes from prior opinions, so that he does not repeat any unnecessary workup and has all of your information prior to surgery.

In The End – *Your* Decision!

In short, this is not a decision that can be made for any patient. The surgeon's job is to evaluate whether or not the patient is a candidate for surgery and to present the patient's options to them. It is a decision that each patient must be comfortable with.

Although it is natural to have some anxiety before any major surgery, every patient needs to have a good understanding of what is involved and be satisfied that they have reached the point where they are willing to commit to the necessary hospitalization, rehabilitation, and potential risks. The best advice here is to take as much time as you need to make a good, informed choice.

Key Points For This Chapter:

- Joint replacements, resurfacings, and arthroscopy are elective procedures.

- Most elective hip and knee surgeries are performed for pain relief when conservative measures are no longer enough.

- Take family & friends with you to the office visit. Privacy laws make it difficult for your doctor to discuss your case with them over the telephone afterwards.

- If surgery is delayed, bones do get softer and muscles do get weaker from disuse of the arthritic joint. However, it is most important to be medically and psychologically ready before deciding it is time to have surgery.

- If there is any doubt, get a second opinion! Most hip and knee surgeries are not emergencies and you have time to get as many opinions as you need to feel comfortable.

- No one can make the decision for you. A good orthopaedic surgeon presents all of your options and information so that you can make a good choice.

Chapter 27 - Scheduling Surgery and What to Expect With Preoperative Testing

Most patients take as much time as they need to decide on if and when to undergo hip or knee surgery (the exception being the patients who need emergency surgery, as for trauma). After thinking about it, reading up on it, and discussing it with family and friends, patients who undergo joint surgery are usually comfortable with their decision, although most are understandably at least a little anxious about the surgery itself.

Scheduling Surgery

The next step is to contact the surgeon's office and let them know you are ready to schedule surgery. In our office, that usually means contacting your surgeon's medical assistant and letting her know that you want to schedule your surgery. She will check the medical chart to see what type of surgery was discussed and help you to pick a date for your surgery.

For joint replacement procedures, it is important to select a time when you can devote about 2 months to recovering. Although you will be up and about, getting out of the house, and driving by 4 to 6 weeks in most cases (and sometimes even sooner), it still takes a couple of months before most patients feel up to taking long trips or returning to physical work. Typical lead times in our practice (and for most busy joint replacement practices) usually are at least 4 to 6 weeks from the date patients call to schedule surgery (sometimes longer for busier times of the year – for example, many patients schedule surgery in January months in advance, and those surgery spots fill far in advance).

It turns out that several weeks are usually needed for preparation, and the usual lead time works well. There are several things that need to be accomplished before the actual hospitalization, and these are explained at the time of surgical scheduling.

Medical Clearance / Preoperative Evaluation

Because hip or knee replacement is an elective surgery, there is time to make sure that patients are in the best health as possible prior to the surgery. For this reason, in our practice (and most joint replacement practices) we ask patients to schedule a visit with their primary care provider a few weeks before the surgery to make sure that their heart, lungs, and other systems are in good shape for anesthesia and surgery. This usually entails a quick physical exam, some blood work, and usually an EKG to check heart function.

Many of our patients come from out-of-town or even out-of-state, and in those instances we have them see one of our hospital-based internists (or hospitalists) when they arrive in town for this visit. This ensures that we have adequate medical testing and history available for their other potential medical issues when they are hospitalized, and it also allows for them to be followed by the hospital internist during their hospitalization. For patients who live close enough for their primary care provider to follow them in our hospital during their stay, it is certainly preferable to have their own family physician perform the medical clearance.

Sometimes the medical clearance and preoperative evaluation turns up some issue that can be optimized, or may even have been unknown to the patient before testing, such as a high blood pressure, blood clotting disorder, diabetes, or cardiac arrhythmia. At least a few times per year, the preoperative evaluation discovers the need for a pacemaker or similar potentially life-saving intervention that the patients may not have even known about. Most of these problems are easily correctable, and patients are then optimized prior to surgery to be in the best health possible.

Preadmission Testing (PATs)

There are also tests that need to be performed within a couple of weeks of the hospitalization. These are usually scheduled to coincide with the preoperative medical evaluation, so that the internist or primary care provider has the laboratory work available for their evaluation as well. The usual tests include a CBC (or complete blood count, which checks for many things, but particularly to see if a patient is anemic prior to surgery), coagulation studies (to ensure that the blood clots normally), and baseline metabolic testing (to check for kidney function, electrolytes, and blood glucose levels).

A chest x-ray and an EKG are also frequently obtained.

At our center, we also perform a series of special x-rays during this visit which we use for templating, or determining the sizes of the components used. Frequently we also image the other bones around the joint being replaced, to check for other issues that may complicate the surgery. For example, sometimes an old fracture that has healed or an area of bowing may require additional preoperative planning to correct limb alignment during surgery.

Another important part of preadmission testing is to check for blood type and make sure that there is compatible blood available in the blood bank in case it is needed. Modern crossmatching actually checks far beyond A, B, AB, and O blood types, and usually a good match can be found before surgery in case it is needed. Many patients prefer to donate their own blood in the weeks before surgery, which can then be given back to them instead of donor blood. Still other patients may not be willing to accept any blood (e.g., Jehovah's Witnesses) and special arrangements are made. This entire topic of blood donation/management merits its own discussion and is discussed in the next chapter.

Joint Class

At our hospital, the preadmission testing, special x-rays, and a short class on joint replacement all occur about 2 weeks before surgery.

The joint class is optional, but I have yet to have any patients tell me they regretted going to it. The class is about an hour long, with separate classes focusing on hip replacement surgeries and knee replacement surgeries. The class is usually taught by physician assistants, nurses, and/or physical therapists, and it serves to discuss the hospitalization, surgery, and issues after the surgery. (As you know since you are holding this book, there is actually a lot to know, and the class helps most people to absorb some of the volume of information that we have been discussing here.) Again, it is a good idea to take along any friends or family members who are interested so that they can hear the class information as well.

Short Term Rehabilitation Planning

Many patients undergoing routine hip or knee replacement surgery will be going directly home, often with arrangements made from the hospital for visiting nurses and/or physical therapists. However, some patients will benefit from going to a short term rehabilitation center for a short while after leaving the hospital and before going home.

The most common reasons for going to a rehabilitation center are social; e.g., a patient who lives alone or does not have adequate help at home during the day may need the extra help at a rehabilitation center before graduating to home. Other reasons are physical, such as having physical limitations (poor stamina or strength, significant obesity, neurologic problems, or other problems) that would make going

directly home difficult. Patients having bilateral surgeries (e.g., both hips or both knees at the same time) also may benefit from short term rehabilitation.

In these cases, we will often discuss the option of short term rehabilitation placement with a patient before surgery and hospitalization. Since there are usually at least several weeks between the time surgery is scheduled and the actual hospitalization, I often recommend that patients visit one or more rehabilitation facilities near their home and pre-book with the facility if they find it to their liking. Most rehabilitation centers are glad to schedule tours and informative meetings before the surgery so that patients can decide if a particular facility is right for them.

Work and FMLA

Most patients have already considered that they will need a month or two away from work before deciding to have surgery, but it is worth thinking about before the hospitalization. Some employers have human resources directors available who can be quite helpful in explaining their company's policies for extended leave and short term disability. Some are able to plan on accommodations such as a temporary desk job or light duty to allow an earlier return to work.

Sometimes family members will want to take time off from work to care for a spouse or parent undergoing surgery. Many employers can facilitate this, with what is called the Family and Medical Leave Act (FMLA). After requesting time off under FMLA, employees may often need to submit some additional paperwork and may need to have the patient's surgeon sign employer forms. It is a good idea to explore this before the surgery and hospitalization if it is needed.

Consent

One final item that is important is the informed consent for surgery. In our state, as in most of the country, the state law requires some form of documented informed consent. Not only is it the law, but it probably is a very good idea.

The consent form for most hospitals essentially describes what the procedure is that you are having done, what location is involved (e.g., right knee, both hips, etc.), name of the surgeon(s) performing the surgery, and that you understand the risks and benefits of the surgery. It is never possible to fully explain all possible risks, so most surgeons list the general and most frequent complications that can occur. Other items on the consent form vary from hospital to hospital, but there usually is a provision to check off as to whether you would or would not be willing to accept a blood transfusion in the event that you need one.

In our practice, the consent forms are usually either signed in the office or mailed to the patient so that they can sign it and mail it back to us. If a patient has not signed the consent before the morning of surgery, then we will usually go over the consent form again at the hospital that morning.

Key Points For This Chapter:

- Once patients schedule surgery, there are usually several things that need to be done in the weeks before surgery

- Medical clearance is usually obtained prior to surgery from the patient's primary care provider or (at our center) from the hospital-based internist for patients arriving from out-of-town

- Medical clearance is a preoperative evaluation to ensure that patients are in the best health possible before elective surgery

- There are some lab tests prior to surgery (usually a blood count or CBC, coagulation studies, metabolic screening, blood type / matching, chest x-ray, and electrocardiogram/EKG)

- Joint class is an optional class before surgery to meet some of the other members of the team such as physical therapists, nurses, and physician assistants and learn more about the surgery, hospitalization, and aftercare

- Some patients will benefit from short term rehabilitation at a facility after leaving the hospital. It is a good idea to discuss this with your surgeon before the hospitalization and visit rehabilitation centers to see which one is right for you.

- Informed consent needs to be signed in the office or at home if the form is mailed to the patient before surgery. If not, it will need to be signed on the morning of surgery.

Chapter 28 - Blood Donation Before Total Joint Surgery

Although not all patients require a blood transfusion with total joint replacement or resurfacing surgery, there is always a chance that you might need a blood transfusion during or after the surgery (curently 3% of primary joints in our practice).

We make every effort to reduce or eliminate the need for blood transfusion. Blood pressure is lowered during surgery, for example, and we use various cautery devices to eliminate bleeding during the surgery. Incisions are smaller than they were in the past, and minimally invasive techniques have led to a reduction in blood loss. For large, revision surgeries or complex surgeries in which significant blood loss is anticipated, a capture circuit is utilized in which blood suctioned from the field is centrifuged and given back to the patient. Some patients may be placed on iron for a few weeks after surgery, which also increases the body's production of red blood cells. This is usually discontinued a few weeks later since a common side effect of the iron is constipation. Vitamin C and folic acid also help with blood cell production.

For many patients, a transfusion may not even be needed for primary (first-time) joint replacements after all of the above measures are taken. However, having a revision or complex surgery, or bilateral surgeries (e.g., both knees or hips at the same time), increases the odds of requiring a transfusion. It is also more important to transfuse older patients or patients with a known history of heart disease or stroke, since it is the red blood cells that carry oxygen to the tissues, and these patients are at higher risk for complications if they remain very anemic. Younger and healthier patients, as a rule, usually can better tolerate a low hemoglobin (blood count) without the need for blood.

For some patients, we know ahead of time that there will be no transfusion, even in the event of an emergency (e.g., Jehovah's Witnesses who have documented their refusal to accept blood products). In those instances, we may use medications preoperatively like erythropoeitin which stimulates the body's own mechanisms to produce more red blood cells. We do not use these methods routinely, however, because there is some concern of side effects and because these treatments are very expensive.

Blood Bank – Donated Blood

If blood is required, there are two sources that are usually employed. The first is the blood bank. Blood samples taken during preadmission testing are matched to samples donated by blood donors; in this way, a sample can be matched very closely. Given enough time, even difficult matches for rare blood types can usually be made by obtaining blood from other banks throughout the state. (Some people are difficult matches because of antibodies, usually generated from multiple previous transfusions.)

The blood bank today is quite safe. Twenty years ago there was much concern because of infections that were transmitted, mainly hepatitis and HIV (which causes AIDS). For the past two decades, tests have been developed and deployed to extensively test blood samples. As a result, the odds of HIV infection via the blood bank are often estimated in the range of 1 in a million or better. Hepatitis is more common.

The principal problem with receiving donated blood today is the small chance of an imperfect match and a transfusion reaction. In most cases, this simply results in a mild fever and the transfusion is stopped, but rarely it can cause a more serious reaction.

The second problem is that of availability; today, the blood supply is dependent on the good will of donors at churches, schools, and work place blood drives. At some times of the year, blood shortages can occur in which transfusions are rationed by need. It is very rare for a shortage to result in someone not getting the blood they need for life-threatening conditions, but we may not be able to fully transfuse someone to desired hemoglobin levels, resulting in their being fatigued for a few weeks until the body regenerates red blood cells.

Autologous Donation

An alternative to receiving donated blood is autologous (self) donation. Patients have the option to donate their own blood a few weeks before surgery if they are in good enough health. This has the

advantage of always being available and having essentially zero risk of a transfusion reaction or infection.

However, there are some disadvantages to self donation. The principal problem is that it does actually make patients more anemic prior to surgery, and for older or less healthy patients, they make not fully make it up before surgery. Studies have shown that patients are more likely to be anemic after surgery if they donate beforehand, and the red cells may not function fully.

There is not a set age limit to self donation, although I generally recommend against it for my own patients over the age of 75, preferring the blood bank instead because it leaves them anemic. However, some surgeons recommend it for all ages. In the end, the option is left up to the patient as long as you are not too anemic (most surgeons use a cut-off hemoglobin level of 11 gm/dl).

If you donate your own blood, usually one or two units of blood (if you have an adequate supply without being anemic already) are taken and stored in a liquid state. The blood is then good for 6 weeks. While it is possible to actually freeze the blood and store it longer, this is not typically done because it requires that the blood is frozen at the time of donation (e.g., it cannot be frozen if it was previously stored in liquid state, as in the case when surgery will unexpectedly be delayed for some other temporary medical issue such as a pneumonia).

Directed Donation

Patients often ask if a relative or spouse can donate blood for them. The short answer is yes, it is possible, but we rarely use it in practice because it still must be matched and go through the same processing that other donated blood goes through. It is also screened for transmittable viruses the same as any other donated blood. Additionally, even a close family member may not be a compatible match. This process typically takes days or weeks, but it is possible.

If a patient is interested in directed donation, we refer them to the American Red Cross for arrangements and will use the tagged blood.

Key Points For This Chapter:

- Despite advances in minimally invasive surgery, use of electrocautery, and other measures, there is still a significant chance that a blood transfusion may be recommended

- The likelihood of needing a transfusion increases with the complexity/size of the surgery (e.g., revision surgeries or bilateral surgeries) and the health of the patient (e.g., already anemic or history of cardiac disease)

- Some patients (such as Jehovah's Witnesses) may refuse blood transfusions, and there are some alternative measures that can be taken such as using erythropoeitin preoperatively

- Patients can donate their own blood before surgery if they are healthy enough

- Donor blood is also available through the blood bank, and it is safe to use but very rarely can result in infection or transfusion reactions

- Directed donation from a friend or family member still requires the same screening and processing process as any other donated blood

Chapter 29 - The Team Taking Care of You

There are significantly more professionals involved in the surgery, hospitalization, and aftercare than just the surgeon. In fact, unless you have gone through the process before, you may be surprised how many different people are needed to make things run smoothly. This chapter takes a moment to list the cast and crew, so that you can be familiar with some of the roles that each member plays.

Orthopaedic Surgeon

The orthopaedic surgeon is the main physician involved during hospitalization for hip and knee surgery. The surgeon is the one responsible for evaluating the need for surgery and performing the surgery itself. The surgeon also manages the orthopaedic care after the surgery in the hospital and in the office in the months afterwards.

Most orthopaedic surgeons have a fairly extensive educational background and training, including 4 years of medical school beyond college, at least 5 years of residency training, and most of those performing joint replacement as a subspecialty have at least one more year of fellowship training in adult reconstruction. Board certification then requires at least a couple of more years of qualified practice and credentialing.

Internist / Primary Care Provider

While the orthopaedic surgeon is focused on the surgery itself and directing the preoperative and postoperative care, most surgeons rely on the patient's primary care provider or the hospital internist (hospitalist) to assist with preoperative evaluation and medical clearance. At our center (and most facilities where a large number of joint replacement surgeries are performed), the internist usually also follows the patient throughout the hospital stay in order to manage blood pressure medications, diabetes, and other nonsurgical issues.

We prefer for the patient's own family primary care provider to follow them in the hospital, but many of our patients come from other cities or states for their surgeries. In these cases, we recommend that patients allow us to consult one of the hospital-based internists (or hospitalists) to follow them during their hospital stay. This internist then transfers care back to the primary care provider after the hospitalization.

Anesthesiologist

The anesthesiologist is the physician responsible for administering and safely monitoring anesthesia during the surgery. For most lower extremity surgeries like knee and hip replacements, they administer spinals, epidurals, or regional blocks in addition to general anesthesia. They usually meet with patients the morning of surgery to explain which type of anesthesia will be used, review the patient's history and medications, and if necessary set up special monitoring like arterial lines, Swan-Ganz catheters, and other special needs. They also monitor the patient's overall status during surgery and in the recovery room, adjusting blood pressure, giving medications as needed, and administering blood transfusions if required. The anesthesiologist usually checks on patients after surgery as well to ensure that they are not having any problems after the anesthesia.

Physician Assistants (PA's)

Physician assistants, or PA's, are professionals who have extensive medical training, including at least 3 years of medical training after an undergraduate degree. They practice under the supervision and license of the physician. In our practice and in the hospital, they are critical members of the team who are invaluable in helping to care for patients. During the day, the orthopaedic surgeon cannot always be everywhere; patients may come to the office for an urgent problem while we are in surgery, or patients may arrive in the emergency room while we are at the office seeing patients. In these cases, we rely on the PA's to be our eyes and hands when we cannot physically be there, communicating back to us what is going on and providing limited treatment (such as treating a hip dislocation or setting a fracture in the emergency room). The PA's also help out in the office, frequently giving joint injections and seeing patients for routine wound checks or suture removal after surgery.

The physician assistants also round on the patients every day, change dressings, order medications, and monitor laboratory and test results. They are usually the first available if there is an issue and we are occupied in the operating room. In our practice, each surgeon often sees all of his patients in the hospital most days, with the physician assistant communicating to us on the days when we cannot be there.

Resident Physicians

In most teaching and specialty hospitals, resident physicians may play a role in caring for patients. At our facility, we often have orthopaedic surgery residents from Yale University (with which we are affiliated) and abroad who come to learn advanced surgical techniques. For this reason, most centers that offer the newest and most cutting-edge technologies and surgeries will often have resident physicians. These doctors have completed medical school and typically at least several years of orthopaedic surgery training at this point. While they may be present as assistants during surgery and in rounding on patients on the wards, they are present to learn, not to perform independent surgery. Patients sometimes express concerns that the resident may be the one to "do the surgery," but in our practice, the surgery itself is **always** performed by the patient's surgeon with one to three assistants present.

Nurses

It goes without saying that the orthopaedic nurses are essential to care and recovery after surgery. At our hospital (and most busy centers that do a large volume of orthopaedic surgery), the orthopaedic unit has its own nurses who are familiar with orthopaedic patients and surgeries. There are nurses who work specifically in the recovery room and those who take over on the orthopaedic ward once the patient arrives.

Physical Therapists

Physical therapists are another essential team member, usually meeting with most patients the day of or the day after surgery. They are trained to help patients safely get up and start moving, and they are responsible for teaching patients how to walk, sit, and get around after surgery. The physical therapist usually provides reinforcement and repeated teaching of the surgeon's directions, particularly in regards to what precautions need to be followed, how much weight the patient can put on the limb, and practicing activities of daily living like putting on clothes and shoes.

Most patients will continue physical therapy after discharge, either at a rehabilitation facility or at home with a visiting therapist, and then often with an outpatient physical therapy facility.

Discharge Planner/Coordinator

Most patients leaving the hospital have special needs for aftercare, either in the form of arrangements for visiting nurses and physical therapists at home or for making arrangements for transfer to a short term rehabilitation facility. At our hospital, we have several discharge planners whose full time job is making these arrangements. They typically meet with patients the day after surgery to begin making these arrangements, and to make sure that everything is in place when the patient gets home. If a hospital bed or special supplies are needed, the discharge planner checks with the insurance carrier and makes sure that these things are arranged.

Medical Assistants

The medical assistant in the office is the person responsible for coordinating and scheduling surgery, preoperative testing, and preoperative medical clearance. In my office, this is frequently the person that most of my patients get to know the best and usually talk to first whenever they have a question or need to get in touch with me. She can get urgent messages to me in the operating room between cases, schedule visits and surgeries, and direct medical questions to either myself, our physician assistants, or other physicians (such as the patient's primary care provider or other specialists, such as a cardiologist).

Visiting Nurse (VNA)

The visiting nurse and/or physical therapist is responsible for seeing the patient at home. Not all patients require VNA services, but many do until they are able to get out of the house on their own. The VNA checks to see that the surgical site is healing properly, looks over medications, does a brief physical exam, and contacts the surgeon's office if there are any problems or concerns.

Key Points For This Chapter:

- The orthopaedic surgeon is the principal physician responsible for care during and after surgery, focused primarily on the surgery and rehabilitation

- An internist (or hospitalist) usually provides medical clearance / preoperative evaluation and follows along in the hospital for all nonsurgical issues (e.g., blood pressure medications, diabetes, etc.)

- The anesthesiologist is responsible for the general anesthesia, spinal, epidural, or regional block and also provides close monitoring and medical management during surgery

- The physician assistants (PA's) are invaluable members of the team, rounding on patients, changing dressings, checking laboratory studies, etc. They are also helpful when the surgeon cannot be present (e.g., in the office or emergency department while the surgeon is in the operating room)

- The orthopaedic nurses are essential for caring for the patient after surgery and are familiar with orthopaedic and surgical issues

- The physical therapist helps the patient get up and start moving safely after surgery and assists in teaching with the proper precautions after surgery

- The discharge planner at the hospital arranges for home needs and equipment as well as transfer to short term rehabilitation centers

- The medical assistant at the office coordinates and schedules surgeries, office visits, and presurgical medical clearance. She is often the first point of contact for patients

- The visiting nurse/therapist sees and monitors the patient at home until they are able to get out of the house on their own.

Chapter 30 - Hospital Admission And Your Medications Before Surgery

Most patients arrive on the morning of surgery at the hospital. Years ago, patients were admitted the day or two before surgery for preoperative testing, but in modern practice most of the testing is performed as an outpatient well before admission.

Preadmission testing also has the advantage of identifying any potential problems – such as a clotting disorder or cardiac arrhythmia – long before the patient has to actually reach the hospital, and consequently, if any minor issues are found they can often be corrected before the scheduled surgery without delaying or canceling it.

There are a few things to remember before the hospital admission, though.

Stopping Medications Before Surgery

Certain medications have to be discontinued prior to admission to the hospital. The most common of these are nonsteroidal anti-inflammatory medications, such as ibuprofen or naprosyn. These medications can thin the blood and should be discontinued **at least 3 days** before surgery.

Aspirin or any drugs that contain aspirin (Anacin, Percodan, etc.) should be stopped **7 days** prior to surgery. These drugs inhibit platelet function so that bleeding is more likely.

Drugs taken expressly as "blood thinners" should be stopped before surgery, but this needs to be checked during the preoperative medical clearance visit. Some of these cannot be stopped (depending on why they are being taken – for example, patients with some cardiac arrhythmias or mechanical heart valves cannot simply stop anticoagulation), and special arrangements may need to be made for admission to the hospital several days before surgery for conversion to a heparin I.V. that can be shut off a few hours before surgery. Warfarin and other "blood thinners" such as clopodigrel (e.g., Plavix) should be stopped **7 days** before surgery if this is cleared with your internist.

Another class of drugs that can cause problems with anesthesia are oral hypoglycemic drugs taken by non-insulin dependent diabetics (such as glucophage or metformin). These should not be taken the day of surgery.

Herbal Medications

Many patients today are taking a number of herbal and nonpharmaceutical remedies for everything from prostate problems to depression. Because these are herbal preparations and not prescribed, many patients may not think to list them when reporting medications to their doctors. However, some of these have potent effects on blood pressure and the nervous system, and can cause potentially serious interactions with some of the anesthetic drugs used. If you are taking any of these preparations, be sure to mention them to all of the physicians involved in your care, including your surgeon, primary care provider performing the presurgical medical clearance visit, and particularly the anesthesiologist.

The Night Before Surgery

Most patients are understandably somewhat anxious the night before surgery. However, it may help to reflect that this is the last night living with that arthritic joint, and after the planned surgery you can begin getting back to life with improved mobility.

Many of our patients may be coming from far away, and it is not uncommon to stay in a local hotel the night prior to surgery. It does, however, necessitate an early checkout.

It is important not to eat or drink anything after midnight the evening before surgery. This ensures that the stomach is empty by the time of surgery. Even for patients just having regional or spinal anesthesia, it is still important that the stomach be empty just in case general anesthesia is needed.

The exception to this is some regular medications you may take such as blood pressure medication. Your internist may recommend taking these medications on the morning of surgery. In this case, it is acceptable to take the medications with **just a sip of water**, as this is a very small volume.

Do not smoke or drink alcohol 48 hours before surgery, as this can cause problems with anesthesia.

What to Bring To The Hospital

Most patients pack a small overnight bag to take with them to the hospital. This is checked in the preoperative area and sent up to the hospital room, along with clothing and other belongings, so that it is waiting for you when you get upstairs after surgery.

The hospital will provide a (admittedly not-so-stylish) hospital gown, which you need to wear for surgery, but you certainly can take along a comfortable gown for after surgery. Pajamas are usually difficult to get in and out of for the first few days after surgery, and most patients find the hospital gowns are more comfortable.

The hospital provides non-skid booties, but a good pair of bedroom slippers are usually helpful to have. You may also want to bring along your own pillow.

The hospital also will provide toiletries, but as with staying at a hotel, you probably will prefer to have your own. Women may want to take a make-up kit, but it is advisable not to wear make-up, jewelry, or nail polish (which interferes with oxygen sensors clipped to the fingers) on the day of surgery.

Reading material and portable music players are fine to bring to the hospital. Most patients also bring a cell phone, which in non-ICU areas is usually allowable, and many patients bring lap-top computers also.

If you have already been given crutches or a walker, bring them to the hospital. Some insurance plans deliver these items to patients' homes prior to surgery. If not, the physical therapist will outfit you with the necessary equipment during your hospital stay.

Be sure to take along any forms and papers given to you by your doctors prior to surgery, especially any consent forms and lists of medications and dosages. You should also bring along a current insurance card(s) in case the hospital needs it.

In general, do *not* bring your own medications, however. It can cause confusion, and the hospital generally provides all medications. On rare occasions, there are some uncommon medications that your internist may recommend you take with you, however, because they may be nonformulary medications (rarely used and not available through the hospital pharmacy). In this case, take the medication bottles (with their labels) in a plastic bag, so that they can be tagged and kept with the other post-operative medications. You will get them back, but it allows the nurses to keep track of exactly what medications (and potential interactions) you are taking.

Also, it is generally not advisable to take valuables, large quantities of cash, or credit cards to the hospital. If you normally wear contacts, plan on using your glasses for a few days instead. It is also advisable to remove all rings (including wedding bands) before going to the hospital, as there can sometimes be swelling from an I.V. or fluid retention that can cause problems with rings.

Key Points For This Chapter:

- Anti-inflammatory (NSAIDs) medications should be stopped at least 3 days before surgery

- Blood thinners such as warfarin, aspirin, or plavix should be stopped 7 days before surgery under the direction of your physician. Some blood thinners cannot be stopped and special arrangements may need to be made.

- Let your doctors know if you are taking any herbal preparations, as some can have potent and dangerous effects with anesthesia

- Do not eat or drink anything after midnight the night before surgery. If your internist recommends taking certain medications (e.g., blood pressure medications) the morning of surgery, these can be taken with just a sip of water

- Do not smoke or drink alcohol 48 hours before surgery

- Do take a small overnight bag with toiletries, slippers, a gown or robe if you want your own, a favorite pillow if desired, and reading materials and/or portable music players

- Do take along any consent forms or documents given to you by your doctor, insurance cards, and lists of medications/dosages

- Do *not* take jewelry, credit cards, large amounts of cash, or your own medications to the hospital

Chapter 31 - What To Expect The Morning Of Surgery

Patients usually arrive on the morning of the scheduled surgery, usually two or three hours before the scheduled time of surgery in order to allow enough time for check-in, getting an IV started, and checking over necessary preoperative lab work and information. The process is similar for all types of joint surgeries, including joint replacement and resurfacing.

Early Admission

Some patients have to be admitted a day or two before surgery because of special medical conditions. The most common cause for this is usually anticoagulation that cannot be stopped (such as for a patient with a mechanical heart valve, who must have the blood thinned at all times). In that case, the patient arrives and is placed on heparin while the warfarin wears off, and then the heparin is shut off a few hours before the surgery.

Patients who require admission prior to the day of surgery typically have a serious co-existing medical condition that necessitates the early admission. This typically also requires prior approval from their insurance company for the early admission.

Same Day Admissions

The majority of elective hip and knee surgery patients arrive the morning of surgery. After checking into the surgery area, a nurse will meet with you and escort you to the preop holding area. Friends and family can visit with you there up until it is time to actually go to surgery.

In the holding area, the nurse will help you get changed into a hospital gown and collect your belongings. These are usually rounded up and placed into bags so that clothing and belongings will be waiting for you upstairs in your room after surgery. It is generally a good idea to leave any valuables, such as rings, watches, or jewelry, at home or with a family member.

Dentures are collected and placed into a container right before surgery, so that they will not be lost and also because it is important that dentures be removed so that they will not interfere with anesthesia (even if you are having a spinal, in case intubation is needed).

If you normally wear contacts, leave them out on the day of surgery and use your glasses instead.

It is also a good idea not to wear any nail polish, because it can interfere with oxygen monitors that may be fitted over your finger.

Please do not shave or wax any surgical sites prior to surgery. Some patients try to do this before arriving, and it actually increases the risk of infection through tiny breaks in the skin. I have had to cancel surgeries when patients (trying to be helpful) waxed their surgical site the day before surgery. If necessary, the nurse will use electric clippers for the surgical area. The side that is being operated on usually will also be painted with an anesthetic scrub.

Preoperative Holding Area Interviews

The nurse will have a short interview with you to go over medications and any recent changes in your health. Do not be surprised if multiple people ask you why you are here and what joint is being operated on; it is not that they forgot or did not know, but there are several check-points scripted in to double-check on the surgery and surgical site. In fact, at our hospitals we ask all patients to mark the surgery site themselves with a marker just prior to surgery.

Next the anesthesiologist will meet with you to discuss the anesthesia and answer questions. The anesthesiologist will have some questions as well in a short interview. In general, we prefer spinal anesthesia if possible for joint replacement surgeries as it is safer and usually more comfortable for the patient, and this is discussed in the next chapter. As the anesthesiologist will explain, a spinal does not mean that you have to be awake for the procedure, and there is almost no chance of neurologic injury from most spinals despite many misconceptions that exist about spinals and epidurals.

Finally, the surgeon comes by to answer any last minute questions and check to see that your health has not changed since you were last seen. If the surgical consent form has not yet been signed, this will be filled out at this time. It is a good time to ask any questions that you might have remembered or formed in the day or two before surgery.

Surgery

After everything has been checked several times and patients are ready (including a quick trip to the bathroom in the preop area if needed), patients are then taken to the operating room on a gurney.

Most patients are surprised that the operating room is such a bright place. Contrary to what is sometimes portrayed on television, the operating room is usually very brightly lit. Patients slide over to the operating table and warm blankets are usually used to keep them comfortable. If general anesthesia is being used, or if a patient does not want to be awake during the time of the surgery, a sedative is given through the I.V. If you have never had intravenous sedation, it is quite rapid, and most patients are next aware of waking up in the recovery room with the surgery complete. It is a good idea to try to relax and think of something pleasant rather than try to fight the sedative, and drifting off is usually quick and pleasant.

If a spinal is being performed, patients may be asked to sit up briefly or lie on their side while the spinal is administered. Again, some patients request medication before this step and often have little memory of it, unless they want to remain awake. The spinal is typically no worse than getting an I.V. placed, and the skin in that area is usually anesthetized with a local anesthetic first.

If a urinary catheter is to be used, it is typically placed in the operating room after anesthesia is established, so most patients do not feel or remember catheter placement. In our practice, catheters are usually only used for bilateral or revision surgeries.

The time required for a particular surgery varies widely depending on the surgeon, surgical approach used, and the size of the patient. Some joint replacements require an hour or less in thin patients, and complex joint replacements, revision surgeries, or those in obese patients may take significantly longer. We are usually back in the operating room longer than the duration of the surgery for other additional reasons, to allow 20 minutes or so for the anesthesia and similar time at the end of the case to get to the recovery room.

Note that surgery may not always begin as scheduled. Sometimes surgeries in progress may run late, and at a busy hospital that also handles trauma, surgeries are sometimes "bumped" if a more urgent case requires use of the next available operating room. Sometimes the anesthesiologist may order an additional blood test the morning of surgery, and this can sometimes cause a short delay as well.

Recovery Room

After surgery, there is a short stay in the recovery room (also called the PACU, or post-anesthesia care unit) until you are fully awake and blood pressure and other vitals are stable. We generally also wait until toes and feet start moving, to signify that the spinal is wearing off. In our practice, all of our patients get x-rays in the recovery room to check on the hip or knee replacement and ensure that there are no problems before starting weightbearing in the evening or next morning.

Usually there cannot be visitors in the recovery room area, but visitors are certainly welcome once you get upstairs to your room.

There is usually some discomfort the first evening. It is expected after major surgery, but there are pain medications available, and you should ask your nurse if you need them. Most patients undergoing joint replacement use intravenous pain medications (such as morphine) the first evening in order to remain reasonably comfortable, and the majority are taking oral pain medications the next day.

Most joint replacement and resurfacing patients can walk and bear weight as soon as the spinal wears off, but it is important to have your nurse or a physical therapist help you the first day or two so that you do not get dizzy or fall. Each time you get up, it gets progressively easier and you will find you can walk farther.

Key Points For This Chapter:

- **Most patients arrive at the hospital on the morning of surgery.**

- **Uncommonly, patients may need to be admitted a few days earlier because of specific medical issues (such as heparin for blood thinners that cannot be stopped).**

- **Patients change and have IV's placed in the preoperative holding area.**

- **In the preoperative holding area, there are quick interviews with the nurse, anesthesiologist, and surgeon.**

- **Patients go from the preoperative holding area to the operating room for surgery, and then to the recovery room.**

- **Patients usually stay in the recovery room until they are awake, moving their extremities, and have had x-rays.**

Chapter 32 - Anesthesia

The anesthesiologist meets with patients on the morning of surgery to explain anesthetic options, risks, and benefits. There are multiple different types of anesthesia that are used, and sometimes a combination may even be employed, depending on many factors such as the patient and their health, what type of anesthesia is needed for the surgery, how deep the anesthesia needs to be, whether muscle relaxation is also needed, and other variables.

For all types of anesthesia, modern techniques and monitoring are typically very safe, with the risks of major complications often being quoted as one in ten thousand or more (depending on the literature source you are quoting). Many patients still have some fear of anesthesia because of risks that were present in the early days of anesthesia. In fact, most of the patients that do have problems while under anesthesia tend to have them not because of the anesthesia, but because the patients are frequently ill and undergoing major surgery for heart, bowel, kidney, vascular, or other problems. Orthopaedic patients, on the other hand, are usually in reasonably good health and additionally are optimized before elective surgery (usually with a complete medical clearance and evaluation prior to large elective surgeries), so anesthesia complications are even less frequent with orthopaedic surgeries.

This chapter discusses the principal types of anesthesia used for most hip and knee surgeries. The majority of our patients have spinal anesthesia for joint replacements, and either regional or local anesthesia with arthroscopic procedures (with or without mild sedation). General anesthesia is sometimes needed for medical reasons or patient preference. At other hospitals across the country, slightly different methods may be used (e.g., all general anesthesia, epidurals instead of spinals for joint surgeries, etc.) and vary by region.

Spinal Anesthesia

Spinal anesthesia is a safe and effective means of anesthesia during surgery, and this is the method most commonly used for most joint replacement surgeries. It involves an injection of medication (usually a local anesthetic) into the lower portion of the spine which numbs and paralyzes the lower half of the body (usually below the navel). Many patients are initially disturbed by this concept, but in fact, spinal anesthesia is usually significantly more pleasant and safer than general anesthesia, with fewer side effects.

Most spinal injections are quick and delivered with a numbing anesthetic (lidocaine) around the skin. The injection itself can contain several different medications, depending on how long the anesthesia needs to last. The injection is usually made at the lower end of the spine below where the spinal cord itself ends; this is important, because it is very rare to have any sort of spinal cord injury for this reason.

Advantages of Spinal Anesthesia

There are a number of advantages to spinal anesthesia. It is safer for patients with any sort of pulmonary (lung) disease, since a breathing tube is not needed. Without the need for an airway, there are fewer risks of airway complications such as obstruction or aspiration of stomach contents.

Spinal anesthesia usually produces excellent muscle relaxation for lower limb surgery, which is important when ligamentous balancing and work on the hip or knee is needed.

Blood loss is usually less for the same operation performed with spinal anesthesia rather than with general anesthesia. This is because of circulatory effects that result in lower blood pressure and heart rate, and thus there is less bleeding at the operative site. Post-operative deep venous thromboses (clots) and pulmonary emboli are also usually decreased with spinal surgery.

Normal gut function also rapidly returns after spinal anesthesia because peristalsis, or the motion of the bowels, continues through spinal anesthesia.

It is important to know that having a spinal does not necessarily mean that you will be awake for surgery (unless you want to be). Most patients elect to have a mild sedative with the spinal so that they doze off for the surgery, then awaken on arriving in the recovery room. However, this level of sedation is mild and does not entail the deep anesthesia of general anesthesia that requires a breathing tube. Patients are usually not as groggy after surgery and have significantly less issues with nausea. It is also more pleasant because it gradually wears off, allowing patients to get situated in their bed and room before fully wearing off.

Disadvantages of Spinal Anesthesia

There are some disadvantages to spinal anesthesia despite all of the above advantages. For the right patient, it is a very good method of anesthesia, but not all patients are good candidates for a spinal anesthesia. Sometimes it can be difficult to get the spinal in if a patient has a very arthritic spine or has had previous spine surgery.

There is a small chance of a spinal headache. This occurs because of leakage of spinal fluid. It usually resolves with lying flat, caffeine, and time, but sometimes a blood patch (injection of the patient's own blood into the spinal cord to form a small clot – it is not as bad as it sounds) is needed to resolve the headache.

Spinals are usually only effective for surgeries lasting several hours or less. Beyond that, general anesthesia is usually needed. This is usually more than enough time for most hip and knee replacement surgeries with experienced joint surgeons, but revision surgeries and complex surgeries are usually planned to last longer and may necessitate other anesthesia arrangements. Sometimes the spinal anesthesia needs to be supplemented with general anesthesia if not fully effective or if it begins to wear off.

Spinal anesthesia sometimes cannot be used if the patient has a condition known as aortic stenosis (narrowing of the aortic valve and outflow region of the heart). This is because the normal blood pressure drop with spinal anesthesia can be severe and problematic if they do not have a sustained after-load (systemic blood pressure).

Spinal anesthesia cannot be used if there are clotting disorders or if the patient has recently (within 24 hours) been on major blood thinners, as this causes some risk of a hematoma and bleeding within the spinal cord area.

Finally, there cannot be any sores or ulcers in the region of the back where the injection must pass through (sometimes a problem for nursing home patients with hip fractures).

Epidural Anesthesia

Epidural anesthesia is similar to spinal anesthesia, except that the injection is given into the epidural space, positioned in the soft tissues just behind the spinal cord. Because of this, it usually takes longer to take effect, but also may not be as likely to produce a drop in blood pressure as a true spinal block.

Sometimes a tiny catheter is introduced into the epidural space and left in place, similar to an IV. This allows for continuous anesthesia for as long as needed, sometimes even days, by continuing to inject small amounts of anesthetic into the epidural space.

Epidural anesthesia essentially has the same advantages and disadvantages of a spinal, except that it can be a sort of continuous spinal that can be used for a prolonged period rather than a single shot lasting a few hours. It also employs a larger volume of anesthetic, and thus it is possible that there can be some complications if this is injected into the wrong space (in a vein, it can possibly cause convulsions or rarely cardiac arrest). While some hospitals use this technique routinely for joint replacement surgeries, we typically prefer the use of a spinal because it does not leave the patient tethered to an epidural catheter. The epidural also needs to be discontinued as post-operative anticoagulants are started.

General Anesthesia

General anesthesia refers to the patient going "completely under," typically requiring placement of a breathing tube and having the ventilator machine breath for them while they are asleep. The term "general" applies because it affects the entire body, with loss of consciousness and motionlessness.

The general anesthetic itself may come in several forms; some are gases that are mixed with oxygen and delivered via a breathing tube or mask (e.g., isofluorane or other volatiles). Other general anesthetics are administered intravenously (e.g., propofol).

For most patients, the last thing remembered is the medication going into the I.V. The breathing tube is typically inserted after sedation and before awakening, so you generally do not remember the tube (although you might have a sore throat afterwards). Because of the use of the breathing tube, there are some risks introduced with this, including loose or chipped teeth and problems with the airway or with the tube being improperly positioned.

General anesthesia can be used for many hours if necessary, and it is needed for very large and complex surgeries. However, it is more likely to produce nausea and vomiting after surgery, and most patients feel drowsy or weak for several days after the anesthesia (like a hangover). It is also more likely to produce mental status changes after surgery in elderly patients.

Regional or Local Anesthesia

These forms of anesthesia are usually used for smaller, outpatient procedures such as knee arthroscopy. Regional anesthesia can be accomplished by several methods, such as femoral and sciatic nerve blocks which specifically block the nerves in the leg, and can also be used in knee replacement cases.

Local anesthesia means that anesthetic (lidocaine or a similar agent) is simply injected into the area being worked on.

Most simple knee arthroscopy procedures can actually be performed with just a local anesthetic injection. In fact, I have had patients who opted to stay awake with the local anesthetic and watch the video monitor with great interest and enthusiasm. However, in the course of each year, there are usually only a few takers with this option (usually only two or three out of hundreds of patients).

Most patients undergoing arthroscopy instead opt for light sedation with the local injection, which usually proves adequate. Not only is it a safe and effective means of rapid anesthesia, but the local anesthetic frequently lasts for several hours, allowing them to get home comfortably and situated with the knee iced and elevated.

Conscious Sedation

This type of anesthesia refers to very light sedation that lasts just a few minutes, typically used in the emergency room for setting fractures, reducing dislocated hips, and manipulating stiff knees under quick anesthetic.

The patient actually remains awake enough to breath on their own but is unconscious enough that they do not remember setting the fracture or performing the procedure. It is usually performed in the emergency room or recovery room, usually with one physician (often an anesthesiologist or emergency room physician) giving the sedation, watching the patient, and monitoring for signs of oversedation while the orthopaedic surgeon does his work. There is a small chance that some patients can become oversedated and may have to be given reversal agents and oxygen via a hand bag until they awaken again, and a very small percentage of patients may require intubation if they stop breathing on their own. However, it is generally safe and performed hundreds of times every day across the country in emergency rooms.

General Considerations

There are a number of options available for anesthesia with hip and knee surgery, and the type selected for a particular patient and surgery depends on the type and anatomical location of the surgery, the length of the surgery, patient factors and other medical issues that may affect the choice of anesthesia, and patient preference.

Typically, most joint replacement and resurfacing procedures will use spinal or epidural anesthesia and secondarily general anesthesia if the anesthesiologist and/or surgical team deems it to be a better choice for the patient and surgery. Knee arthroscopy usually uses regional or light general anesthesia, sometimes being performed with only local anesthetic.

Regardless of the type of anesthesia used, there are a few important things to remember. When discussing anesthesia with the anesthesiologist, it is a good idea to mention any allergies, problems with medications in the past, and any prior experiences with anesthesia. It is also important to mention all medications normally taken as well as to inform the anesthesiologist of any herbal supplements you may be taking.

If you drink alcohol on a regular (i.e., daily) basis, or certainly if you use any recreational substances such as marijuana, it is vital to be honest and let the anesthesiologist know this. The information is kept in confidence, and a potential reaction can be very serious.

It is also important to let the anesthesiologist know if any blood relatives have had bad anesthesia reactions in the past, and if you have any loose teeth or dentures. Finally, if you have had anything to eat or drink since the evening before, be sure to let the anesthesiologist know.

Key Points For This Chapter:

- Modern anesthesia is safe and serious complications are rare.

- Most orthopaedic patients are undergoing elective surgery and are typically optimized and medically cleared before large surgeries, further decreasing the risk.

- Spinal anesthesia numbs and paralyzes the body below the level of the navel for a few hours. It is safe and has a number of advantages over general anesthesia for patients who are good candidates.

- Spinal anesthesia does not mean you necessarily have to be awake for the surgery (most patients also opt for a mild sedation).

- Epidural anesthesia is similar to spinal anesthesia, but has the option of leaving a small catheter in place for delivery of anesthesia for hours or potentially even days.

- General anesthesia refers to going "completely under," usually requiring a ventilator and breathing machine.

- General anesthesia is more likely to result in nausea, vomiting, and "hangover," but it is sometimes needed for prolonged complex surgeries.

- Regional and local anesthesia are useful for arthroscopy and other minor (usually outpatient) surgeries

- Conscious sedation is very quick anesthesia that lasts for minutes, usually used for setting fractures or dislocated hips in the emergency room

- The anesthesiologist, surgeon, and you will determine the type of anesthesia that is best for you based on your medical issues, health, type and length of surgery, and any specific surgical issues that need to be addressed.

- Let the anesthesiologist know about all medications, allergies, prior anesthesia, family history of reactions to anesthesia, herbal supplements, recreational drug use or regular alcohol use, loose teeth or dental problems, and when you last ate/drank anything.

Chapter 33 - What to Expect After Joint Replacement: Getting Around, Physical Therapy, Medications

Patients usually arrive upstairs on the orthopaedic ward within a few hours after finishing surgery and their brief stay in the recovery room. Although many of our patients who have had their surgery early in the morning may be up and about in the afternoon or evening, some other patients rest the first afternoon and evening after surgery. Pain medication is given when patients request it, and ice is applied to the surgical site. For most patients the first night is uncomfortable but tolerable with pain medications and antiemetic (anti-nausea) medications. After that, each day gets easier.

Day 2 (The Day After Surgery)

On the morning of the second day, physical therapy begins. In fact, we often have patients up the evening of surgery if they feel up to it and the spinal has worn off enough. The most important thing after any joint replacement surgery is to get up and moving as quickly as possible. The faster patients are up and about, the faster they feel better, and mobilizing also helps with other problems like preventing blood clots and constipation.

Hip replacements have some precautions with certain movements for the first couple of months after surgery in order to prevent dislocation. In particular, you may be told not flex the hip beyond 90° or cross the legs until six weeks or so after the surgery (by which time the tissues have healed and tightened up around the joint). In contrast, anterior hip resurfacings do not have any range of motion limitations or precautions in most circumstances, because these devices are very difficult to dislocate. Knee replacement patients do not have precautions, but they do need to learn how to properly begin walking and moving. Physical therapy will reiterate the precautions that apply to your surgery so that you don't forget.

Lab Tests

Lab tests are usually drawn the morning after surgery, and one of the things that is checked is the hemoglobin and hematocrit. These are measures of anemia post-operatively and are used to help determine if someone may need a blood transfusion or not. Clinically, it is usually fairly straightforward to determine who clearly needs a transfusion as these patients usually get quite lightheaded when they are up and about. Other tests usually include basic metabolic panels and electrolytes, and coagulation studies are sometimes checked depending on the method of anticoagulation being used.

All patients are on some form of blood thinner (anticoagulation) to prevent blood clots after hip replacement surgery (the next chapter is dedicated to discussion of blood thinners and clots prevention). In our practice, most first-time hip replacements (e.g., not revisions or hip fractures) will be started on enteric coated aspirin twice per day unless they are at higher risk because of smoking, a history of clotting problems, or bilateral surgeries, in which case low molecular weight heparin (lovenox) may be used. If patients were on warfarin before surgery (commonly for cardiac reasons, such as atrial fibrillation), then they can resume that after surgery. Note that different surgeons may utilize different blood thinners, and there is a wide variation on which methods are used.

Physical Therapy

During this first day of physical therapy, we are principally concerned with teaching patients how to start exercising and doing basic activities (such as using a commode) on their own. Every patient will be different in terms of how fast and how far they can walk; it depends greatly on the physical condition of the patient before surgery, how extensive the surgery was, how anemic the patient is, what additional medical problems the patient may have (e.g., lung disease, morbid obesity, etc.), and other factors.

In general, we prefer for patients to walk at least some short distance. Frequent breaks are used. As patients get more proficient and confident with their

new hip(s), they eventually can begin getting around on their own using a walker or cane, but initially these activities need to be supervised by nurses and physical therapists.

Most patients are very apprehensive about getting up for the first time. The discomfort after surgery however is usually due mostly to the muscle and soft tissue pain from the incision, and the majority of patients are surprised to find that the replaced hip or knee joint actually feels fine with weightbearing. There is no longer the grinding, arthritic pain of a worn out joint. The incisional pain gets dramatically better after the first couple of days, and this can be helped significantly with ice packs or cryotherapy units.

Diet

Usually most patients can begin at least with clear liquids within a few hours after surgery. Diet will gradually be resumed, but as a general rule of thumb, it is a good idea to start slowly with bland things like toast, crackers, and jello before progressing to meat and potatoes!

If you feel nauseous, be sure to let your nurse know. There are medications (like compazine or odansetron) to counter-act nausea.

The pain medications used after surgery often decrease appetite. Many patients may be eager to start a diet after getting a joint replacement, but this is not the time. Dieting can begin in a few months, after tissues have finished healing and you are exercising again. Even if you do not feel like eating very much, it is important to keep drinking plenty of fluids. The intravenous fluids can usually be discontinued once you are taking plenty of liquids orally.

Although we do not use warfarin as a routine blood thinner in our practice, some joint replacement centers prefer it for all of their patients for post-operative thromboembolic prophylaxis. In this case, certain foods that contain a significant amount of vitamin K should be avoided, and a nutritionist will usually help educate you about what foods can interact with the warfarin (collard greens, spinach, sweet potatoes, and others). The vitamin K acts as a natural antidote for the warfarin and cancels its effectiveness.

Dressings and Drains

In our practice, we typically try to remove dressings and leave the incisions exposed to air on the day after surgery. Sometimes there is still significant drainage (which can sometimes last for days, particularly for obese patients) which may necessitate the use of a dressing. In general wounds fare better when open to air than with perspiration and a warm, moist dressing without air. However, surgeons in other practices may have different instructions about when to remove the dressings.

Do not let anyone touch your incision without washing their hands or using antiseptic gels. Most hospital staff know this by heart, but it never hurts to be vigilant because of the risk of infection. After all, it is *your* incision.

Some redness and swelling is normal and to be expected. It does not mean there is an infection. Bruising is also not unusual.

Drains and catheters are usually removed on the first day after surgery (not all patients will have drains or urinary catheters, particularly those undergoing short, first time replacement surgeries). Both may be left for longer if there are medical reasons to do so. As soon as patients are taking good oral intake of fluids at least and do not require transfusions or other IV access, the IV will be discontinued.

Discharge Planning

Usually on the day after surgery one of the discharge planners (or social worker in some hospitals) will meet with you to discuss making arrangements for aftercare. For patients who are going to a short term rehabilitation facility, the discharge planner either makes the arrangements or confirms them if you have already prebooked with a rehabilitation center. They also help to make arrangements for visiting nurses and physical therapists at home and can also help with straightening out any issues with insurance carriers.

Days 3 and 4

Physical therapy and getting up continue to get easier. Again depending on the patient and the surgery, some will begin practicing with a cane and others will continue to use walkers or crutches. The most important factor is simply getting up and moving.

We are watching at this point to be sure the incision is healing properly, there are no medical complications (blood clots, ileus, etc.), and that patients are eating and voiding properly.

It is normal to be constipated for a few days after surgery, and the laxatives and stool softeners that patients are given will help. One of the most important influences in return of normal bowel function is how often the patient is up and about after surgery. Patients who had spinal anesthesia will usually have a quicker return to normal bowel function than those who had general anesthesia.

Discharge

At this point, the majority of first time joint replacement and resurfacing patients will be ready to go home on the third or fourth day in the hospital, using visiting nurse (VNA) and home physical therapy visits until they are ready to go to outpatient therapy. Some patients will go to a short term rehabilitation center, typically those who have had more complex surgery, had bilateral (e.g., both sides) surgery, have other medical issues, or those that may live by themselves or in a home situation in which they do not have adequate assistance.

Most patients are able to travel in a regular car. Moreover, most insurance plans do not cover an ambulance ride to home or to the rehabilitation facility unless there is a clearly defined (and documented) medical reason for this. The discharge planners who arrange for home physical therapy, visiting nurses, and transfers to a rehabilitation facility can arrange for medical transport, but unfortunately they cannot make your insurance carrier pay for the services. Some patients do elect to obtain these services at their own expense if they feel they need medical transport.

Patients are discharged not only with their own discharge instructions, but also extensive instructions for either the visiting nurse/physical therapist or for the rehabilitation facility. These include lists of medications and dosages, weight bearing status, activities to do and to avoid, problems to telephone us about, when to follow up in the office (usually 3 to 6 weeks in our practice, depending on the procedure and surgeon), and other instructions.

Key Points For This Chapter:

- **Most first time joint replacements and resurfacings require 3 or 4 days in the hospital.**

- **Most first time joint replacement surgeries take between 60 and 90 minutes per side, depending on patient size and other factors.**

- **Patients usually arrive the day of surgery unless they have a medical problem that necessitates earlier admission (such as blood thinners that cannot be discontinued).**

- **Patients are up and walking within 24 hours of the surgery.**

- **Each day, mobilizing and getting up are the most important factors. It gets progressively easier.**

- **Most patients go home with home visiting nurses/physical therapy, but some go to rehabilitation facilities depending on age, general physical condition, and social factors.**

Chapter 34 - Preventing Thromboembolism After Joint Replacement And Use Of Anticoagulation

A thrombus is a clot, and a pulmonary embolism is a thrombus that breaks loose and travels through the veins, eventually reaching the lungs where it can cause a serious blockage of the blood flow going through the lungs and back to the heart. Thromboembolic disease refers to both of these related problems.

A pulmonary embolus is the most serious result of thromboembolic disease, but other problems include painful and persistent swelling and circulation problems if the valves in the veins are damaged from blood build-up behind the clot, leading to chronic venous hypertension.

Although thromboembolic disease is uncommon in the approximately 800,000 patients undergoing total joint replacements in the U.S. each year, it is one of the major risks and complications that can occur with major orthopaedic surgery of any sort to the lower extremities. Not surprisingly, there has been much research and effort to reduce or eliminate thromboembolic disease in the 40+ year history of joint replacement surgery. Today the risk of getting a symptomatic deep vein clot while on prophylaxis is around 1% to 5% (although silent, asymptomatic clots are probably somewhat more prevalent).

Deep Vein Thrombosis (DVT)

A deep vein thrombosis is the formation of a thrombus (or blood clot) within the deep veins, usually in the calf or thigh. There is a difference between superficial phlebitis and deep clots; the clot that forms within the very large veins in the lower extremity tends to be the one that causes concern. This clot can cause problems by itself, or it can break free and cause a pulmonary embolism.

The thrombus (clot) can partially or completely occlude the blood flow. This usually leads to significant swelling below the level of the clot, similar to a back-up in the plumbing system, because blood flow back to the heart is impeded by the clot. Nearly all patients have some degree of lower extremity swelling for months after major joint surgery, but severe swelling and calf or thigh pain are reasons for potential concern.

Some DVTs are not symptomatic. In fact, some studies have suggested that there are many more silent DVTs than are diagnosed because many do not cause any problems and resolve on their own. However, it is better to prevent their formation in the first place than to try to treat them once they occur.

Factors Contributing To DVT

Deep vein thromboses (clots) can occur even without surgery. This can happen after a prolonged trip (such as a transatlantic airline flight) or in anyone who is bedridden for a prolonged period of time.

Stagnant blood flow through the veins increases the possibility of a clot forming. This happens with prolonged bed rest or immobility. Not surprisingly, the best prevention for blood clots is early mobility and getting up as quickly as possible after surgery.

Length of surgery appears to also play a role. Although it is most important to do a good job with the surgery itself, this is one of the reasons it is better to have a surgeon whose operating times are shorter and quicker. (Incidentally, there is probably also an increased risk of infection with longer operating times.)

Coagulation, or clot formation, is a complex process which the body normally uses to prevent us from bleeding to death from cuts or injuries. In normal situations and even surgery, this normal coagulation cascade or process is beneficial. However, some patients may be prone to forming clots too easily by having underlying medical or genetic conditions that lead to hypercoaguability. Diseases like antiphospholipid syndrome, protein S or C deficiency, or a host of other genetic diseases can increase the likelihood of forming clots.

Obesity, diabetes, varicose veins, and cancer all have been shown to increase the risk of forming clots. Smoking and use of oral contraceptives also increase the likelihood of forming clots.

Injury to veins can cause clots. In particular, "kinking" of the vessels during surgery from either retractors or by awkward positioning of the leg may increase the likelihood of clot formation. There is significant evidence that surgical approaches that do not necessitate twisting the leg during surgery have less risk of clot formation; this is one of the reasons we favor the anterior approach for hip surgery in which the patient is laying flat (supine) without the need for twisting the leg during surgery (as is the case for most posterior approach surgeries).

Incidence

There is a wide variance between published reports regarding the incidence of DVT in orthopaedic surgery patients after joint replacement. One principal reason for this is the large number of asymptomatic clots that are only detected if screening tests are performed, as in research studies. Most hospitals do not routinely perform tests to check for clots (e.g., ultrasound or venograms) unless there are symptoms to suggest such a test is needed.

Some studies suggest that the incidence of DVT in patients without any prophylaxis after total joint replacement may be in the range of 40% to 80%. With prophylactic treatment, most studies report the incidence of clinically symptomatic thromboembolic disease to be less than 5%.

The most important figure is probably the number of fatal pulmonary embolisms, since this is the outcome that we are most concerned with. Most studies report that incidence to be well under 1% in patients treated with some form of prophylaxis. In our practice, we have published studies looking at the incidence of DVT and PE after hip replacement. In a series of over 2000 total hip replacements, we found about 1% incidence of clinically significant and diagnosed DVT and there was only one death from a pulmonary embolus out of several thousand patients.

Tests To Detect DVT or PE

Most surgeons and hospitals do not screen for DVT with routine diagnostic tests, but instead order these tests if a patient demonstrates clinical symptoms. The most common symptoms of a DVT are swelling in the lower extremity, edema, increasing discomfort (instead of decreasing discomfort in the time after surgery), and certain physical examination signs such as palpable cords in the back of the knee or a positive Homan's test.

The most common signs of a pulmonary embolus are sudden chest pain and shortness of breath. A patient may require supplemental oxygen to keep blood oxygen at adequate levels.

Both pulmonary emboli and deep venous thromboses often cause a mild elevation in temperature. However, it is common for patients to have a low grade fever (less than 101.5° F) for a few days after major joint replacement surgery from other factors as the body adjusts after surgery, and this is not a specific finding.

A venogram is probably the most accurate test to diagnose a deep blood clot, but it is not often used because the test itself may increase the likelihood of a clot and requires injection of contrast into the vein. Therefore, it is not routinely used in most institutions to diagnose blood clots.

An ultrasound (doppler venous ultrasound) is probably the most common test for detecting blood clots. This is a painless test in which an ultrasonic probe is placed against the back of the leg and used to view the deep blood vessels and visualize any clots that may be present.

Less commonly employed tests include magnetic resonance imaging (MRI) or plethysmography. MRI is helpful for diagnosing clots deep within the pelvis, where it is difficult to detect clots with an ultrasound, but it is a very expensive test and is not always readily available in smaller hospitals. In addition, not all patients can have an MRI (such as pacemaker patients) because of the magnetic field. Plethysmography is an older test that uses blood pressure cuffs at different locations along the leg to detect blockages, but few centers use it today because it is not as accurate as more modern methods.

A pulmonary embolus may be diagnosed with a VQ scan (also known as a ventilation-perfusion scan) or a special CT scan with contrast. A VQ scan is a nuclear imaging test in which very slightly radioactive substances are taken in to see if ventilation (breathing) matches the perfusion (or blood flow in and out of the lungs). If not, there may be a pulmonary embolus. A more accurate test is to perform a spiral computed tomography (CT) scan with a contrast agent in the blood to see if there is a clot situated within the pulmonary artery that leads away from the heart to the lungs. However, patients with kidney disease cannot have the contrast dye, and a VQ scan is used instead for these patients. An arterial blood gas sample may often be obtained (in contrast to most blood draws, which are from veins) to analyze the blood for oxygen and carbon dioxide levels, which can also be helpful in diagnosing a pulmonary embolus.

Prevention

The best treatment for thromboembolic disease is to prevent it in the first place. As mentioned before, the best prevention occurs from getting up and moving as quickly as possible. In addition to walking, patients are advised to move the ankles up and down frequently throughout the day to keep the circulation moving.

The form of anesthesia used during surgery can have some bearing on the risk for DVT formation. The rate of DVT formation can be decreased by up to 50% in some studies by using regional or spinal anesthesia instead of general anesthesia.

There are mechanical forms of prophylaxis, such as stockings, pneumatic pumps, and intravenous filters, as well as pharmacologic forms such as aspirin, heparin, or warfarin. Most surgeons use a combination of both, but there is no clear evidence that any particular method of pharmacologic prophylaxis is significantly better than others and as a result, there is wide variation in what type of prophylaxis is used. The highest risk period is probably the first 5 days, although there is evidence that there is a second peak in incidence about 2 weeks after surgery. It is also not clear how long prophylaxis needs to be used, although most studies suggest it is probably best employed for four to six weeks after surgery (possibly longer if a patient is not getting up and mobilizing well).

Mechanical Prophylaxis

Some patients may be too sedentary or forgetful to remember to keep moving their legs and ankles, which is why most hospitals also use some form of mechanical prophylaxis for most joint replacement patients. This usually is in the form of a pneumatic sleeve that fits around either the calf or the foot and periodically inflates and deflates, moving the circulation throughout the leg.

Another method of mechanical prophylaxis is the use of compression elastic stockings, which encourage venous blood flow back to the heart. These are more effective in preventing blood clots in the calf than the thigh.

The most aggressive mechanical prophylaxis is to place a small filter in the vena cava, the large vein that returns blood flow to the heart. These filters resemble a small umbrella and are usually made out of wire. These are placed in the vena cava by a vascular surgeon or radiologist via a catheterization, and while some types of filter are designed to be temporary, the majority are designed to be left in place permanently. The idea is that the filter will catch any large clots before they reach the heart, trapping them in the wire arms and holding the clot in the circulation until it breaks down. On the downside, these filters can sometimes migrate and travel to the heart or other locations. For this reason, filters are not usually used unless a patient has a high risk based on risk factors or a history of previous thromboembolic disease or blood clots.

Aspirin

Aspirin is probably the simplest and least expensive pharmacologic method of DVT prophylaxis. It is easy to take (unlike heparins that require injection), does not require monitoring (as with warfarin, which requires periodic blood tests), and does not usually lead to bleeding complications (e.g., hematoma). However, some patients cannot tolerate aspirin because of gastrointestinal issues or ulcers, interaction with other medications, or kidney disease.

We routinely use aspirin for most straight-forward, single-sided, first time total hip replacement patients in our practice. It is not the strongest blood thinner, however, and is not used for high risk patients, most knee replacement patients (who have a higher risk of blood clots in the calf), or patients having extensive or bilateral surgery. The usual dosage for prophylaxis after surgery is 325 mg of enteric coated aspirin twice a day for a total of 650 mg of aspirin daily (many patients may already take a "baby aspirin" each day for their heart, but this is a smaller dosage at 81 mg once a day).

Heparin and Low Molecular Weight Heparins

Heparin is a naturally occurring substance that inhibits clotting. It is useful because it has a very short period of effectiveness (hours), which allows it to be used for patients who may need the blood thinner "turned off" to allow a return to the operating room for additional procedures. It is also used for patients who have a condition in which the blood thinner can only be discontinued for a very short period of time to allow surgery (such as patients with mechanical heart valves who have to be on blood thinners every day of their lives).

Heparin only comes in injectable forms, commonly used as a subcutaneous injection in the soft fat of the abdomen for prophylaxis after surgery. Not all patients are able to give themselves the injections.

Heparin has been used for many years in its high (unfractionated) molecular weight form, but in recent years low (fractionated heparin) molecular weight heparin injections have become available and are in widespread use (such as enoxaparin). These low molecular weight heparins only need to be injected once or twice a day (instead of every few hours) and work faster than aspirin or warfarin, but are very expensive (often costing hundreds of dollars, as compared to the cost of aspirin each day). In addition, there is some strong evidence that these blood thinners are more likely than aspirin or warfarin to cause post-operative bleeding complications (such as a hematoma), which can sometimes be serious.

Warfarin

Warfarin (which is often known under the brand name of coumadin) has been used for decades as a blood thinner. It was originally developed as a rat poison (named after the Wisconsin Alumni Research Foundation where it was developed over 50 years ago), and works by interfering with vitamin K metabolism and the blood clotting cascade. It is taken orally and does not require injection. It also does not usually cause gastrointestinal or kidney problems, but it does have some downsides of its own.

Warfarin works slowly, often taking days to reach its desired effect. It also requires frequent monitoring with blood tests to make sure the blood is not too little or too greatly thinned (problems can arise from both). Although the medication is inexpensive, the laboratory testing and monitoring each week is not.

Because the natural antidote to warfarin is vitamin K, eating foods rich in vitamin K can decrease the effectiveness of the medication. Many medications can interact with warfarin, such as cholesterol medications, some antibiotics, and some herbal medications.

Warfarin can result in hemorrhage, excessive bruising, bleeding from the nose or gums, and blood in urine or stool. Although rare, a specific complication known as warfarin necrosis can occur that can be devastating and fatal (leading to massive thrombosis in the skin and limbs, with problems of necrosis and gangrene in the limbs). We very rarely use it in our practice, but across the country many surgeons still commonly use it as an accepted practice.

Alternative Blood Thinners

There is currently a great deal of interest in developing alternative pharmacologic agents for DVT prophylaxis because of the hundreds of thousands of patients each year who need it and the various downsides of the major methods currently used (aspirin, heparins, or warfarin).

There is interest in newer antiplatelet drugs such as clopidegrel, but thus far there is not sufficient evidence that this is adequate for DVT prophylaxis. Several promising anticoagulation drugs in development in recent years were subsequently shown to have unexpected side effects and development has slowed or stopped with these medications. Some alternative anticoagulants are in development, but these will not likely be ready for widespread use for a number of years at the time of this writing.

Treatment of DVT or PE

When a thromboembolic complication occurs (DVT or PE), treatment depends on several factors. Minor, superficial clots do not typically require any treatment beyond continuing the prophylactic dosage of whatever regimen the patient is already on.

Deep venous thromboses in the calf veins usually do not require inpatient heparin treatment and can often be treated on an outpatient basis with either warfarin or low molecular weight heparins for 6 to 12 weeks. Thromboses in the thigh may require bed rest and a week or so of heparin therapy followed by several months of warfarin. Pulmonary emboli of significant size usually require heparin, close observation and oxygen therapy, and subsequent treatment with warfarin for usually at least six months.

Key Points For This Chapter:

- A deep vein thrombosis (DVT) is a clot that forms inside the large veins, usually of the calf or thigh.

- DVT probably occurs in 40-80% of patients without some sort of prophylaxis, although many DVTs may be asymptomatic and resolve on their own.

- A pulmonary embolus occurs when a piece of clot breaks off and travels through the circulation to the lungs, which can be fatal. It occurs in a fraction of 1% of joint replacement patients.

- DVT can be diagnosed with clinical examination (calf swelling or tenderness), ultrasound, venography, or MRI.

- Pulmonary emboli are usually diagnosed with a CT scan or VQ (nuclear ventilation – perfusion) scan.

- Mechanical prophylaxis for blood clots includes early movement and mobilization, exercises, pneumatic pumps, compression stockings, and on rare occasions a wire filter placed in the vena cava (large vein going to the heart).

- Pharmacologic prophylaxis for blood clots commonly includes aspirin, heparin (including fractionated low molecular weight heparins), and warfarin.

- There is a wide variation in accepted medical practice for preventing thromboembolic disease. Different surgeons and hospitals use different methods, and there is little evidence that one particular method is better than another as long as some sort of protection is used.

- There are advantages and disadvantages to each method of pharmacologic prophylaxis.

- Treatment of DVT/PE usually involves being placed on higher levels of anticoagulants and sometimes bedrest.

Chapter 35 - What to Expect After Leaving the Hospital In The First 3 Weeks

The hip and knee sections each discuss the specifics of the weeks after surgery, and there are some differences between each. In general, knee replacements are somewhat more challenging in rehabilitation than hips because there is a greater range of mechanical problems and primarily because the knee will become quite stiff if it is not aggressively moved.

However, there are some generalities for the first three weeks after surgery that are common for all joint replacements (both knees and hips) and bear mentioning.

Incision Care

Most surgeons will allow showering (but not submersion!) when the wound is dry. Some surgeons advise waiting until staples or sutures are removed, although in our practice we generally let patients begin showering 2 days after leaving the hospital if the wound is dry and "glued" without staples. Check with your surgeon about when it is acceptable to begin showering (it is usually in the discharge instructions).

Incisions usually have some redness along the scar for many months. Most knee or hip incisions gradually become a thin white line over the first year after surgery. Some patients are prone to keloid formation, which is harmless but can result in a less cosmetic scar.

Absorbable sutures can sometimes "spit," poking through the skin with a tiny bit of string visible and sometimes some associated fluid. This is not uncommon and is not typically any cause for concern.

Some physicians advocate rubbing vitamin E oil over incisions to decrease scarring. There are also a number of commercial ointments and salves that purport to decrease scarring. There is not significant evidence that these treatments really improve wound healing, although the massaging of the surgical site itself probably is helpful. I often recommend breaking open vitamin E capsules and rubbing the incisions to my patients because the massaging action helps to decrease fibrous scar adhesions.

New incisions are definitely prone to sunburn, however, and if you go to the beach or are out in the sun in the first 12 months after surgery, you should take care to protect the scar from the sun. It should be covered with clothing ideally, or at least very high SPF sunblock.

Swelling

Some degree of leg swelling is normal after hip and knee surgery, and it is not unusual for some patients to even notice a slight difference between the size of the legs for months after surgery. However, it should steadily be improving, and any swelling that suddenly becomes markedly worse should be reported right away as it can be a sign of a blood clot.

Bruising is also normal for a few weeks. This gradually resolves.

Noises

It is common for most joints (hips and knees) to make some noise after surgery, often in the form of clicking or popping. As long as there is no specific pain associated with the noise, it is usually harmless. There are numerous reasons for the noises, which are often from tight ligaments or scar tissue or from the contact of the artificial components themselves. Rarely, some materials (e.g., ceramic total hip replacements) can have some "squeaking" noises. Any noises that are associated with specific pain should be reported.

Fever

It is common for most surgery patients to have mildly elevated temperatures in the week or two after surgery. However, persistent fevers for more than a week or two, or particularly high fevers beyond 101° F, can be indicative of infection and should be called in. It is a good idea to keep a thermometer at home and simply check if you feel any chills or as if you may have a fever.

Drainage from the Wound

Most hip and knee incisions are dry by the time patients leave the hospital, although it is not unusual for some to have drainage for a week or possibly more. It is more common in larger, obese patients or patients who are undernourished (and have slower healing). It is probably also more common in smokers.

A particularly common source of drainage is the site of a drainage tube if one was used. This is usually the last spot to close up.

As long as the drainage is clear, yellow, or just bloody (sanguinous), there is usually not much need for concern although it should be reported to the visiting nurse or surgeon if it persists more than a few days. Drainage that changes character and becomes thick, green, white, or foul-smelling can be indicative of infection and should be reported right away.

Pain

Some soreness after surgery is expected. The analogy that I often tell my patients is that if I kicked them in the shin, I would expect it to be sore for a couple of weeks, and surgery is usually a little more than a kick in the shin.

However, soreness should gradually be getting better week by week. Any sudden changes, particularly severe ones, should be called in. These are often simply the result of muscles resuming normal activity, and most patients have been relatively inactive with the affected hip or knee because of the need for surgery in the first place.

Patients usually are prescribed various medications for pain. Most patients use narcotic medications for the first week or two, and sometimes beyond that as there is a great deal of variation from person to person in pain tolerance. It is preferable to discontinue narcotics as soon as possible, however, as these do have some undesirable effects such as constipation, nausea, occasional confusion, drowsiness, and a tendency to build up a tolerance to the opioid medication over a few weeks. In general, we prefer for our own patients to transition to Tylenol within a couple of weeks if possible.

Numbness or Tingling

Some numbness around the incisional area is normal and is usually the result of tiny cutaneous nerves that are necessarily transected with the skin incision. This usually resolves over a period of weeks to months and is not typically bothersome for most patients, although if it affects a large area, care will be needed with using ice or heating pads to make sure that the skin does not burn.

Numbness that is worsening, on the other hand, needs to be reported. Weakness in bringing the foot upwards at the ankle that is new or worsening should be reported.

General Considerations In The Home

Although most patients are ambulatory by the time they get home, many are surprised by the little things that they may need assistance with during the day. Suddenly preparing meals and taking out the trash are major events for a while. Fortunately, these things soon get easier within days to weeks.

Before surgery, it is a good idea to look through the house and try to anticipate ways to make post-operative life less challenging. Loose area rugs or extension cords that run across the floor should be taken up. A shower stool is helpful, and grab bars in the bathroom are even better. If possible, it is often helpful to stock up on frozen food so that preparing meals is easier.

It is usually helpful to have a cordless phone that you can clip onto your belt, particularly since getting to the phone is likely to take a little longer in the first few weeks. Some patients have found fanny packs or clothes with big pockets to be helpful so that they can keep small items handy.

Most patients after joint surgery can sleep in a normal bed. However, some patients find that it is helpful to order a hospital bed to have it on a downstairs level if they normally sleep upstairs. Many insurance plans will cover a hospital bed for a short period of time if necessary.

Managing Post-operative Visits and Medications

Most surgeons have patients return in 3 to 6 weeks after surgery for a quick wound check. Be sure that you have your appointment arranged after you get home. Like most practices, we recommend that patients call and schedule their post-operative visit as soon as they get home from the hospital.

The hospital sends a list of discharge medications and dosages home with the discharge instructions. The visiting nurse usually has this information also. Most patients do not go home on an antibiotic unless there is a specific reason to do so, but if you have been prescribed one, be sure to take it as directed. Any blood thinners are also very important to take (e.g., aspirin, warfarin, or heparin injections). Most other medications, such as for constipation, nausea, and pain medications, are taken as needed.

In general, most aspects of daily life and function in the hip or knee should steadily improve each week after surgery. Any change in this direction of progress, where something becomes more difficult or uncomfortable rather than steadily improving, is often a sign that you should check with you surgeon's office.

If you have any questions or problems, call your surgeon. Many patients have a visiting nurse that sees them regularly who can also answer questions. Most practices have a medical assistant or physician assistant available to return calls if the surgeon is unavailable or operating that day.

Key Points For This Chapter:

- Follow your surgeon's instructions regarding incision care.

- Some incisional redness, swelling, soreness, and bruising are normal. If any of these get **WORSE** over time, however, call your surgeon.

- Some drainage for a few days is not unusual. Drainage that persists beyond a week or changes character (e.g., thick or foul-smelling) should be reported.

- Noises are commonly heard from the joint after surgery. These are generally harmless unless they are associated with pain.

- Low grade temperature elevations are normal in the first week or so after surgery. However, persistent fevers or temperatures greater than 101° F should be reported.

- Some numbness in the immediate area around a surgical incision is common and not cause for alarm. It is usually due to cutaneous nerves that are transected at the time of surgery, and this usually resolves over a long period of time.

- Try to prepare the home ahead of time with things to make life easier after surgery.

- Be sure to call your surgeon's office if you think you are having a problem after surgery.

Chapter 36 - Complications of Hip and Knee Surgery

When you step onto a modern jet airliner, you typically do not dwell on the tens of thousands of parts that have to work correctly in order to fly you across the country. In fact, if you really knew in detail the thousands of things that *could* go wrong, you might feel daunted about getting on the airliner in the first place. Yet at any given moment, there are thousands of jets in the air, and we rarely hear about crashes.

The same scenario holds true for major surgery. The vast majority of patients undergoing orthopaedic joint surgery every day do quite well, with an estimated 95% to 98% of patients having a good outcome, but it is a complex effort made by a team of professionals working together. Sometimes complications can and do occur despite all of the modern technology and herculean efforts made to prevent them.

For this reason, we always re-iterate in our practice that joint surgery is elective and should not be undertaken until patients are completely willing to accept the risks of surgery in order to get better. For most patients, pain and disability generally reach this point after conservative options no longer provide adequate relief.

This chapter goes into detail about the more common complications that can occur after major orthopaedic joint surgery, but it is not possible to list every potential complication. Rare complications can occur that are outside the scope of this discussion, but complications themselves are uncommon and this list discusses the ones that are seen most often. The complications here are broadly grouped into categories of anesthesia complications, medical complications, and surgical complications.

Anesthesia Complications

Airway Problems

Airway problems can occur if a patient needs to be intubated (i.e., have a breathing tube placed) during surgery but has a swollen or abnormal airway. In severe cases, this may necessitate an emergency tracheotomy. Sometimes the breathing tube may not be properly placed into the airway and is instead in the esophagus, but there are multiple checks to prevent this, including electronic monitoring of carbon dioxide (which should be seen with exhalation) and listening with a stethoscope. A mild sore throat may result from intubation, although most patients do not recall the intubation because the tube is placed after sedation and removed before awakening.

Dental Problems

If a patient has loose teeth, these may be further knocked loose during intubation, the process of placing a breathing tube for general anesthesia. Loose teeth should be addressed by a dentist prior to surgery if present. Cracked teeth can rarely occur from the intubation process. Dentures can get in the way, and these should be removed prior to surgery in case intubation is necessary.

Malignant Hyperthermia

This is a rare but potentially serious reaction to some anesthetics in which the patient's body temperature raises rapidly after induction of general anesthesia. It usually is determined by a genetic predisposition that runs in families, so the anesthesiologist may often ask about any prior problems with anesthesia in yourself or any family members. This is not usually a concern with spinal anesthesia.

Spinal Headache

Some patients may have a severe headache after spinal anesthesia. This is because of a pressure change in the cerebrospinal fluid. It is usually not dangerous, but it can be uncomfortable. It is usually treated with caffeine and resting flat. If the headache is severe and prolonged, sometimes an anesthesiologist may administer a blood patch. A blood patch is a spinal injection with a minute quantity of the patient's own blood, in order to produce a clot over the site of spinal fluid leakage. This usually resolves the problem.

Medical Complications

Heart Attack / Stroke

These complications are rare during or right after elective joint surgery, in part because most patients are "tuned up" for surgery and optimized after obtaining presurgical medical clearance. A small percentage of patients may experience these complications as a result of low blood pressure (hypotension) during or right after surgery.

Cardiac Arrhythmia

Some patients can spontaneously develop a cardiac arrhythmia after surgery, often for reasons that are unclear. This may require monitoring in the telemetry unit until the symptoms disappear. Rarely, this may require treatment with either blood thinners or electric cardioversion by the cardiologist.

Congestive Heart Failure (CHF)

Patients who have preexisting heart disease may have decreased pumping ability of the heart (often measured preoperatively by the cardiologist as an ejection fraction). Sometimes the heart has difficulty keeping up with the pumping demands, and fluid begins to back up into the lungs because the heart is not pumping as strongly as it should. This is usually treated with fluid restrictions, diuretics ("water pills"), and cardiac medications prescribed by the cardiologist to help the heart pump more strongly.

Blood Clots (Thromboembolic Disease)

An entire chapter is devoted to this topic, but after major surgery in the lower extremities, clots can form in the deep veins of the leg or pelvis (deep venous thromboses, or DVT). The DVT may cause swelling in the leg and calf that may be uncomfortable, but is not usually dangerous by itself. The principal concern is that a portion of the clot may break away into the blood stream and travel back to the heart, through the right side of the heart, and then to the lungs and lodge there. That event is called a pulmonary embolism, and it can be immediately fatal. In our published series, we documented one case of fatal pulmonary embolism in over 2000 consecutive anterior approach total hip replacements.

Fat Emboli

Fat emboli result when fat from the bone marrow enters the circulation and causes damage to small blood vessels elsewhere. It is not usually serious, but can result in mental status changes and respiratory problems (sometimes major respiratory failure in the case of trauma patients). Rarely, it can cause coagulation problems. Fat embolism is seen more often with fracture care than with elective joint replacement surgeries, particularly with surgeries that require reaming and rodding through the marrow of the long bones.

Disseminated Intravascular Coagulation

Disseminated Intravascular Coagulation (DIC) is a rare complication in which blood starts to coagulate throughout the entire body. Because the body uses up all of the platelets and other materials that it normally uses for clotting, patients paradoxically have simultaneous problems of hemorrhage and clotting at the same time. This potentially fatal condition is usually only seen in critically ill patients and patients who have had massive trauma, very extensive surgeries with massive transfusions, cancer, liver disease, or sepsis (systemic infection in the blood).

Pneumonia / Atelectasis

Small sacs within the lungs (called alveoli) may close down, particularly if a general anesthesia is used instead of a spinal. This is called atelectasis, and by itself this can reduce oxygen exchange and also lead to low-grade fever. This is why deep breathing and use of incentive spirometers is encouraged after surgery. For frail or ill patients, a post-operative pneumonia can occur (this is more of a concern for hip fracture patients than for elective joint replacement patients).

Urinary Retention

Sometimes patients have difficulty voiding on their own after surgery and anesthesia. If this occurs, it usually is the result of anesthesia and narcotics, and resolves shortly on its own but may require catheterization for a day or two. For men, an already enlarged prostate can sometimes be the culprit, and this can sometimes be treated with medications. If it remains a problem after a short period of time, then a urology consultation is usually obtained and the patient may rarely go home with a catheter until the underlying problem has resolved.

Urinary Tract Infection

Many elderly patients, particularly women, may be prone to urinary tract infections even without surgery. However, if a urinary catheter is needed, there is some chance of introducing a urinary tract infection. This is not usually serious but may require antibiotics. For this reason, catheters are discontinued as soon as possible after surgery.

Clostridium Difficile (C. Diff) Colitis

Prolonged use of antibiotics can lead to the destruction of the "good" bacteria that normally inhabit the gut and overgrowth of the "bad" bacteria, namely a particular bacteria called *clostridium difficile*. This usually presents as intractable diarrhea. It can be treated with special antibiotics, but it is best to prevent it by only using antibiotics when needed. It can be fatal, particularly in frail or elderly patients (such as hip fracture patients).

Ileus

It is normal for the gastrointestinal system to slow down for a few days after anesthesia and surgery, and most patients do not have a bowel movement for at least a couple of days. However, sometimes the gut can stop moving entirely, particularly in obese patients and/or those patients taking heavy doses of narcotics, resulting in bloating and abdominal pain that is termed an ileus. This normally resolves by stopping all oral intake of food and liquids, mobilizing out of bed and walking, giving IV fluids to prevent dehydration, and sometimes placing a nasogastric tube to suction that keeps the stomach empty until the condition resolves.

Mental Status Changes

It is not uncommon for elderly patients (particularly the very elderly and patients who already demonstrate some mild evidence of early Alzheimer's dementia) to have post-operative confusion and mental status changes. It is usually worse at night, hence the common term "sundowning." This can last for days, and in uncommon instances, the confusion can last for weeks. It usually resolves as narcotics are discontinued and the patient returns to more familiar surroundings, but it can be distressing for other family members.

Delerium Tremens

The DT's occur as a result of alcohol withdrawal. Heavy drinkers cannot simply stop their usual intake; otherwise, mental status changes and tremors can occur that in some cases can actually be life threatening. Treatment usually involves tranquilizers, folate and potassium, and in severe cases resuming regular dosage with alcohol. It is important to tell your doctor if you drink on a daily basis so that they can be prepared for this possibility and try to prevent it.

Transfusion Reaction

Some patients may react to receiving donated volunteer blood from the blood bank. It is rare to have a serious reaction, although such reactions have been well described. Usually a transfusion reaction results in fever and itching, and the transfusion is simply stopped. Most transfusions are run very slowly, over several hours, so that when a reaction is observed it can be stopped immediately. The symptoms usually resolve with benadryl and fluids.

Hyperglycemia

Blood glucose levels may be elevated in diabetics and pre-diabetics after surgery or trauma for a number of reasons. Some diabetics who normally are controlled with oral medications may require insulin for a short period of time. Some patients who have never been diagnosed as diabetic may require treatment during their hospitalization; it does not necessarily mean that they are diabetic and often can just be a stress reaction. This usually resolves fairly quickly as the body re-adjusts from the stress of surgery.

Electrolyte imbalance

Numerous metabolic disturbances can cause individual electrolytes to become off-balance, namely sodium (hyponatremia or hypernatremia), potassium (hypokalemia or hyperkalemia), magnesium, and other electrolytes. Some of these derangements can potentially cause serious problems such as cardiac arrhythmias, but most are not serious and are corrected as needed by administering or restricting the appropriate element.

Edema

It is normal for the lower extremities to have some degree of swelling after orthopaedic surgery, often for many months. In unusual cases, it can sometimes be quite pronounced. There can be many underlying reasons for this, but it usually resolves with elevation and time. Some male patients who have had very extensive pelvic trauma or surgery may notice scrotal edema, which can be alarming but usually resolves on its own without any significant problems.

Renal Failure

Failure of the kidneys to filter the blood adequately is uncommon in healthy patients, but patients with severely low blood pressure (as seen after a major trauma, for example) or with preexisting renal insufficiency (as with many diabetics) may see significant decreases in the functioning of the kidneys. This is usually temporary. Some medications may contribute to this problem (e.g., nonsteroidal anti-inflammatory medications or NSAIDs) and may need to be discontinued. Rarely, temporary hemodialysis may be needed in severe cases.

Surgical Complications

There is some overlap between medical and surgical complications, but surgical complications are generally related to the incision and joint replacement itself. Some specific risks to each type of surgery are also discussed in the this section.

Hemorrhage

Although very uncommon, with any major surgery there is the risk of uncontrolled bleeding during (or even after) surgery. It does not happen often, but sometimes a blood vessel can be injured and may be difficult to clamp or cauterize, leading to increased blood loss. If this occurs, the outcome is usually the need for a blood transfusion during or after the surgery, but on rare occasions the blood loss can be more serious. Unlike surgeries within the chest or abdomen, however, there are few vessels in the hip or extremities large enough to produce bleeding that is fast enough to be a serious danger, allowing surgeons time to control the bleeding.

Infection

Deep infection of the joint (sepsis) is probably one of the worst surgical complications that can occur. Despite all of the efforts made to prevent it and use of perioperative prophylactic antibiotics, rates of deep infection in joint replacement surgeries still are reported at about 1 per 100 patients.

In some cases, this may necessitate additional surgery to "wash out" the joint, or possibly even remove all artificial components and treat with antibiotics for several months until re-implantation can be considered after clearing the infection. Infection is an increased risk for patients who are obese, smoke, have problems with their immune system, are on immunosuppressive drugs (such as transplant patients or some patients with autoimmune diseases), diabetics, or patients who have had previous surgeries in the same location.

An infection that involves the bone itself is termed osteomyelitis. This usually only develops in chronic infections that have been present for a long time. Many cases of osteomyelitis which do not involve artificial implants can be treated with antibiotics that travel to the bones, but sometimes surgical intervention is required if this fails.

Today there is increasing awareness about MRSA, or methicillin-resistant staphylococcus aureus, a particularly nasty bug that has become problematic in hospitals around the country in the past fifteen years as bacterial resistance to modern antibiotics increases. If a patient has ever had MRSA, in most hospitals they will be kept in private rooms and isolated from other patients because of the potential risk that they may still harbor the resistant bacteria and could pass it to other patients.

Cellulitis

Cellulitis refers to a superficial infection that occurs in the skin, usually a patch of red, warm, tender skin over the surgical site (some redness and warmth is normal, however). This is not usually a serious problem and resolves quickly with treatment, but surgeons take it seriously because it can sometimes spread to deeper tissues or into the joint if left untreated. Most cases of cellulitis are treated with elevation, antibiotics, warm compresses, and marking out the area involved and observing it.

Wound Dehiscence

Dehiscence means a breakdown of the wound. A wound may break down and open up for several reasons, including infection, poor wound healing ability (e.g., a malnourished elderly patient, patient who has had chemotherapy or radiation, or diabetic), or pressure from below the incision due to a fluid collection. Superficial dehiscence is usually not serious, but merits close observation. The wound usually fills in on its own over time, usually a period of weeks. Rarely, it may need to be packed with a gauze dressing or a vacuum dressing until it fills in.

Other Wound Problems (Tape Blisters, Decubitus ulcers, etc.)

Patients can sometimes develop a variety of minor, superficial skin and wound problems. Tape blisters are common in patients with sensitive skin or thin skin (such as elderly patients or patients who have been on prednisone). These usually heal fine without any specific intervention once the dressings are removed.

Pressure sores, also known as decubitus ulcers, can occur if a patient is not getting up much, commonly on the back of the heel or buttocks. The principal treatment is to remove pressure from the affected skin area, by mobilizing, placing gel pads, rolled towels under the heel, and other similar measures.

Suture abscesses or "spitting sutures" are also common and usually harmless, but a patient may notice a tiny bit of exposed suture several weeks after the surgery. This is usually nothing to be concerned about and dissolves on its own.

Hematoma

A hematoma is a collection of blood. It can sometimes be quite large and cause pain and swelling around the surgical site. Some degree of hematoma is normal after all surgeries, but very large ones sometimes need to be drained. In rare cases, the hematoma expands to a large enough size that it can compress nearby nerves or blood vessels. This usually only occurs if a patient is on high dosage anticoagulants (blood thinners).

A related problem is a seroma, which is similar except the fluid collection contains serous (usually joint) fluid rather than blood. Obese patients are more prone to seroma formation as some of the underlying adipose tissue liquefies after surgery.

Sometimes a hematoma can spontaneously decompress on its own days or weeks after surgery. While it can be distressing to a patient who was not expecting it (it appears similar to menstruating from the wound, with old clots expressing from a small opening), it usually stops on its own in a day or two and the patient actually feels better. Usually a dressing is all that is required until the wound closes up again.

Fracture

With most joint replacements and resurfacings, there always exists the possibility of fracturing the bone while placing the implants. This is one reason why we always obtain x-rays in the recovery room before starting physical therapy. This can vary from an incidental finding on x-rays to a serious problem that requires additional surgery or fixation, although when it occurs, most patients are simply treated with limited weightbearing for a few weeks until the fracture has healed. Some patients may not fully fracture until later, as they place weight on a weakened area of bone. It is a greater concern in frailer, older patients (such as elderly hip fracture patients), patients chronically on steroids (prednisone), or in complex revision surgeries.

Nerve Injury

Nerve injuries can occur with any surgery, and the specific nerves involved vary according to the region in which surgery occurs. Any superficial skin incision will often result in localized numbness, which is usually harmless and resolves over a period of months. Sometimes more serious injuries can occur, such as a peroneal nerve palsy which manifests as a foot drop. This is usually seen in posterior approach hip surgeries or when correcting a knee valgus (knock-kneed) deformity and the nerve is stretched. Rarely it can sometimes be seen after knee arthroscopy because of stretching the nerve while positioning the knee.

Foot drops (or peroneal or sciatic nerve palsy) frequently get better on their own but may require many months to do so. Sometimes an orthosis is required to keep the foot up and from dragging the ground.

Anterior approach hip surgery can result in rare injuries to the femoral nerve or the lateral femoral cutaneous nerve (meralgia parasthetica, which affects the skin area over the front of the thigh). Many nerve injuries are the result of stretching (as from retractors) rather than direct injury, a condition known as neuropraxia, and this often resolves over a period of months without the need for any additional intervention.

Vascular Injury

It is also possible to injure a nearby blood vessel with any invasive procedure, although it is rare with hip and knee surgeries. In the hip, one particular area of concern is injury to blood vessels inside the pelvis while placing screws into the socket (acetabulum), which may penetrate large vessels on the other side of the bone. In knee surgeries, the major arteries and vessels are in the very back of the knee, but these can rarely be injured by instruments, saws, or retractors. Major vessel injury in routine knee surgeries is exceedingly rare (it is more common in complex revision surgeries), but it can be a potentially devastating complication that may result in loss of the limb if the vessel damage cannot be repaired.

Heterotopic Ossification

Sometimes the body can form "extra bone" or dense calcified scar tissue in the region of a joint in the months after surgery. This is typically seen around the hip, and it is more common in men and in larger, more complex surgeries. It can sometimes be prevented (at least to some degree) by immediate radiation therapy just before or after surgery (within 72 hours), or by starting medications right after surgery that prevent extra bone formation (such as indomethicin or celecoxib), but these treatments are not usually undertaken unless a patient has a prior history of heterotopic ossification. In some cases, the stiffness may limit range of motion enough to warrant surgery to remove the calcified tissue.

Limp

All hip and knee joint surgery patients limp for at least a while. There can be many reasons for persistent limp, but the most common reason is weakness (atrophy) of the muscles around the affected joint from disuse prior to surgery. The joint can be replaced, but the muscles only get stronger through exercise and therapy. The muscles around hip and knee replacements get stronger for about a year after surgery.

Sciatica and Bursitis

Surgery does not usually cause sciatica or bursitis, but it can aggravate these problems if already present. Low back pain can sometimes flare up also right after surgery. Sometimes the prolonged period of lying down during surgery and convalescence can aggravate these problems. However, these issues usually resolve uneventfully with anti-inflammatory medications and/or physical therapy, just as they are treated in patients who have never had surgery. These problems usually resolve within a few weeks, although some patients who have had hip replacement may come back from time to time to get a steroid injection for bursitis.

These problems may also frequently be seen in patients *before* surgery, often as a result of an altered gait pattern due to a bad hip or knee.

Component Loosening

Over a long enough period of time, all implants will eventually loosen from the bone. Cemented components will generally loosen before noncemented (porous-coated) components. Joints that contain polyethylene (plastic) bearing surfaces are also particularly prone to osteolysis, or reabsorption of the bone. There are a number of other factors, such as patient activity level, weight, and types of activities (e.g., impact activities such as running are more likely to loosen the components). Usually loosening is a slow, gradual process that leads to slowly progressive pain (often in the thigh for femoral loosening and in the groin for acetabular cup loosening) over months to years. A bone scan detects the process before it is visible on x-rays (although a scan will not be useful until at least a year after surgery because of normal bone uptake around a new prosthesis), and patients usually can plan on revision surgeries usually well before the time that loosening becomes debilitating.

Component Wear

A related longevity problem is wear of the bearing surfaces. Over time, the artificial joints simply wear out the surfaces, like brake pads in a car. Ceramic and metal bearings very rarely wear out and usually require decades before significant wear occurs, but polyethylene (plastic) bearings may wear out in the years after surgery. This usually manifests as pain in the affected joint, and often there is associated osteolysis (or re-absorption of the bone around the prosthesis) and loosening because of the wear particles. In fact, it is the loosening that usually necessitates revision surgery, although plastic liners are sometimes replaced in a smaller surgery if the wear is noted on x-rays before osteolysis becomes a problem.

Noises (Pops, Clicks, etc.)

It is not uncommon to experience "noisy" joints after any surgery, particularly joint replacement or resurfacing. In fact, most patients experience popping, clicking, or other noises even before surgery. For the vast majority of patients, it is a harmless phenomenon that is probably more common than not after surgery. There are multiple reasons for noises after surgery.

The ligaments and soft tissues around a joint frequently are the culprits with painless noises. Artificial joints can often produce clicking when the hard surfaces of the components come together. Normal arthritic joints frequently crack and make crunching noises, called crepitus, which can persist after surgery.

Painful noises, on the other hand, may be a cause for further investigation. Tissue or loose bodies that become trapped in the joint, as is the case with the painful clicking from a meniscal tear, may need treatment. Ceramic surfaces have also been reported to uncommonly cause "squeaking" noises.

Complications Specific To Hip Replacements/ Resurfacings

Dislocation

A hip dislocation occurs when the ball slips out of the socket. It is typically quite painful and results in immediate shortening of the leg and inability to bear weight on it, similar to a fracture. It has to be put back into the socket in the emergency room or operating room under anesthesia. Rarely, surgery may be needed to put it back into the socket if there is muscle or other tissue blocking it from going back into place. However, once it is back in the socket, most patients immediately feel better and frequently are able to walk out of the emergency room and go home (with careful precautions so that it does not happen again).

If only one or two dislocations occur, the tissues usually will tighten up over time and no further treatment may be needed. However, if a patient continues to experience multiple dislocations, there may be an underlying problem that needs to be addressed with revision surgery (commonly from an acetabular socket that is too vertical or tilted forwards/backwards, impinging scar tissue or bone that allows the hip replacement to fulcrum out of the socket, increased tissue laxity that has occurred since the original surgery, or most commonly a patient who is noncompliant and does not follow instructions – they may need conversion to a more constrained hip replacement that is more difficult to dislocate).

In our published review of over 2000 hip replacements with the anterior approach, dislocation rates were around 1%. These typically occurred within the first 6 weeks of surgery, before the muscles and tissue has "tightened up" after surgery. For this reason, there are some precautions that we advise total hip replacement patients to follow during this period. As of the time of this writing, we have yet to see any dislocations from hip resurfacing, although it is still technically possible and these surgeries have only been performed on a routine basis for a relatively short period of time.

Ceramic Problems

While ceramics have the potential to last a very long time (decades), there are some downsides other than being very expensive. Ceramic balls and liners can on rare occasions fracture. For this reason, impact activities like running or basketball are not recommended for ceramic bearings. In most of the rare cases we have seen, there has often been a trauma that required equivalent energy to breaking a bone or a manufacturing defect in the ceramic. For most patients, the risk of ceramic fracture is far out-weighed by the benefits of a long-lasting bearing surface.

Ceramics can also be the culprit for audible noises, such as squeaking or clicking. While not usually harmful, these problems can be annoying for the rare patient in which they occur (generally estimated at about 1 in 400 ceramic hip replacements).

Metal on Metal Problems

Metal on metal bearings are another bearing surface that has the potential to last significantly longer than a traditional metal on polyethylene (plastic) joint replacement, but these also have some specific drawbacks. Metal ions accumulate over time in the body, particularly in the lymph nodes, liver, spleen, and other tissues. Metal on metal bearings have been used for decades in tens of thousands of patients, and while increased metal ion levels can be measured in the blood and urine of these patients, there has not to date been convincing evidence that this presents a danger except to those patients who have metal ion allergies or those who have kidney disease (because this is how the metal ions are normally excreted). We typically do advise patients that while these are very tough and long-lasting bearings, there may be a small chance of problems with metal ion accumulation in the future. Women who may become pregnant are also typically advised to consider alternatives such as ceramic hip replacements because of the potential for metal ion accumulation.

Implant failure

Although this is not as much of a problem as it was several decades ago, before the engineering of joint replacement devices advanced to where it is today, on rare occasions the implant itself can break. We typically see this today with breakage of implants that were placed many years ago, and we do not expect to see it as much in the future as high performance implant designs have improved. However, it does usually necessitate revision surgery when it happens.

Leg Length Discrepancy

Nearly all hip replacement patients will at least feel as though one leg is longer than other for several months after surgery. This is because of the muscles on the affected side that are healing after surgery. Sometimes it is because of pelvic obliquity, in which the lower lumbar spine tilts (somewhat like scoliosis, but located at the very lower part of the spine where it meets the pelvis). This results in an apparent leg length difference.

However, true leg differences do occur in hip replacement surgery. Surgeons are careful to try to match the length as much as possible by a number of different techniques during surgery (in our practice, we use an anterior approach with the patient lying flat, so it is simpler to check that the knee caps are at the same level). However, the most frequent reason for a true length discrepancy is stability. At the time of surgery,

the hip is held in the socket by muscle and soft tissues; if there is inadequate tension, the hip may dislocate. Most surgeons test the stability with trial components of different sizes before implanting the final components, and if the hip dislocates too easily, the surgeon must decide if extra length is necessary (there can be other reasons for dislocation, too, such as suboptimal positioning of the acetabular cup). By making the leg a little longer, the tension is increased and the hip is more stable.

Relatively few patients have a length difference significant enough to require a shoe lift, but some do. All patients undergoing hip replacement must be willing to accept that it is a possibility and potential trade-off for hip stability. It is significantly more likely with complex revision surgeries.

Subsidence

The femoral stem can sometimes "sink" further into the upper portion of the femur in the weeks after surgery, leading to shortening of the leg and sometimes resulting in instability or dislocations. This can occur in soft bone or with noncemented prostheses that are not fully seated. This is one of the reasons post-operative x-rays are taken a month or two after surgery.

Thigh Pain

There are several potential causes for thigh pain after hip surgery, with the most common being trochanteric bursitis. Within the first month few months after surgery, occasional thigh pain is normal and expected. Some patients may experience thigh pain in the first 18 to 24 months after surgery, especially after a cementless hip replacement, as the bone grows into the porous coating of the prosthesis. It is usually not constant or severe and resolves once bone growth is adequate.

Occasionally, thigh pain that is severe and worsening over time can be indicative of more serious problems. These can include early loosening, stress fractures, or infections. Fortunately, these problems are quite rare in comparison to the routine thigh pain or bursitis pain above, but for this reason it is important to let your surgeon know if thigh pain develops and persists or worsens in the years after surgery.

Stress Shielding

Not so much a complication as a phenomenon, stress shielding occurs because bone will sometimes reabsorb around an implant if the implant is so mechanically solid and "stiff" that the body decides the bone above it is no longer needed. This is typically seen with certain types of noncemented femoral stems that become very solidly fixed in the thigh, and the body reabsorbs the bone around the upper portion of the stem over a period of years. This is one reason that femoral stems are often made from titanium, which is strong but allows some bending. Other designs have a "clothespin" design at the bottom so that the implant is not so mechanically stiff. Stress shielding is not usually a problem in itself, although it makes revision surgery tougher if the implant ever needs to be taken out and replaced because there is less bone left to work with for reconstruction.

Complications Specific To Knee Replacements

There are a number of problems that are primarily associated with knee replacements and surgeries. These are more likely with joint replacements than with minor procedures such as arthroscopy, and total knee replacements in general are somewhat more likely to have complications than partial knee replacements.

Stiffness (Arthrofibrosis)

Any joint can become stiff after injury or surgery. Generally, elbows and knees are more concerning for this problem, which is why physical therapy starts so soon after surgery and focuses on aggressive range of motion. Sometimes additional manipulations or other interventions are required for stiffness that does not adequately improve with physical therapy. In knee replacement surgeries in particular, a patient who develops significant stiffness (particularly obese patients, diabetics, or patients who are slow to mobilize) may require manipulation under anesthesia in which the scar tissue is broken up by manipulating the knee under light anesthesia for a few minutes. There is a small chance of fracturing the bones or dislodging the prosthesis when manipulating the knee under anesthesia, usually in frail, older patients with osteoporosis. In a few cases, arthrofibrosis may require surgical debridement by arthroscopy or an open procedure (arthrotomy).

Ligamentous Instability

Sometimes a knee can be unstable after surgery. There can be many reasons for this; typically, it is after total knee replacement when one of the ligaments surrounding the knee is either stretched, cut, injured, or improperly balanced. Knee replacements are more difficult to balance than hip replacements, primarily because the joint is more complex. The knee is not a simple hinge mechanism, but actually swings like a four-bar linkage. Additionally, it has some rotational movement as well (the "screw-home" mechanism). Most knee replacement designs rely on keeping some or most of the knee's ligaments intact (except for true hinge designs, but these do not work as well as designs that keep the patient's ligaments). If a knee is initially unstable and gives way in a particular direction (often from side to side), it will usually improve with time as the ligaments heal and readjust to the knee replacement. Sometimes a knee brace may be needed if the knee tends to buckle after surgery.

Extensor Mechanism Disruption

This is one of the more serious mechanical complications of knee replacement. The quadriceps attach to the patella, which in turn attaches to the patella tendon, which is anchored on the front of the tibia. This entire assembly is commonly referred to as the extensor mechanism; if this is disrupted – for instance, the patella tendon pulls away from the tibia during or after surgery – then there is no power to the extensor mechanism. Thus, the knee will have no ability to extend and the quadriceps cannot do its work. This is a serious complication that requires revision surgery to repair the extensor mechanism at whichever point it has failed, sometimes even requiring tendon grafts to strengthen the failed mechanism.

There are times when the knee is so stiff or tight that the extensor mechanism has to be surgically taken apart in order to perform the knee replacement. This is

usually only seen during complex revisions or when there has been previous knee surgery and scarring, but occasionally a stoic patient will put off surgery until the knee is so stiff that this step is required just to perform the knee replacement. In this circumstance, the disruption is planned and surgically repaired at the end of the case, typically by detaching the bony anchor at the tibia and then wiring it back or by cutting the quadriceps tendon and then suturing it back together at the end of the case (I much prefer the former technique as patients seem to recover faster). However, either technique will necessitate walking in a knee immobilizer and not bending the knee for at least 4 to 6 weeks after surgery and result in some residual stiffness.

Balancing Issues

Because the knee does not move as a true hinge, it is a challenge to correct the many different variables that are necessary for optimal stability and ligamentous balancing. Commonly, many knees may be tight in flexion but loose in extension, or loose in flexion but tight in extension. These issues usually arise because of the depth of chamfer cuts on the end of the femur or the depth, slope, and tilt of the tibial bone cuts. Typically these issues, if present, will improve or resolve over time as the knee strengthens and regains mobility.

Varus/Valgus

Varus describes a knee that is "bow-legged," and valgus describes a knee that is "knock-kneed." Both terms are used to describe a leg that is not in anatomical alignment. Most knees prior to surgery have worn one side of the knee away faster than the other, resulting in a bowing of the leg (most of the time, it is in varus, or "bow-legged).

At the time of surgery, we try to correct this angle back to an anatomic alignment. However, this is not always possible to do for a number of reasons. If a patient has a severe angular deformity or bow prior to surgery, chances are good that there will still be at least some deformity after surgery. Sometimes surgery can result in an overcorrection. As with most mechanical complications, patients must be willing to accept that possibility before surgery.

Rotational Problems

Just as the knee can have angular problems with bowing from side to side, it can also have rotational issues. Sometimes the foot is rotated outward or inward ("pigeon-toed") after surgery. This may result after attempting to fix balance and angular problems. It is usually a more cosmetic problem than functional unless the rotation causes problems with the tracking of the patella.

Patella Tracking Problems

The patella, or knee cap, has to glide up and down in the groove over the femur. This can be a problem even in many patients who have never had surgery, and sometimes can lead to the patella jumping out of its track (dislocating) or having pain. Some patients in fact have surgeries to correct this problem (and there are a number of alternative surgeries to try to fix this).

However, this problem can also appear after knee replacement. It is most commonly associated with patellas that have been resurfaced with a plastic button, which may cause the patella to track differently or catch

and "clunk" as it passes over the metal flange of the knee replacement. In some cases, the tracking problem may be severe enough to warrant revision surgery.

Internal Derangement

Some knees have problems with catching and locking after surgery. Typically, this develops months or years later, and is often associated with scar tissue or meniscal remnants that are getting caught in the hinge of the joint. Sometimes it can be caused by a loose body, such as a small bit of cement that has become loose and is floating about in the knee. If symptoms are severe enough, arthroscopic surgery is sometimes used to take a look inside the knee.

Effusions

Some patients can have persistent effusions (or collections of fluid) in the knee joint after surgery. Some effusion is normal for at least a month or two, but persistent, large amounts of fluid that accumulate in the knee may need to be drained. The most common reason for a fluid accumulation is that the patient has been very active and "overdid it," such as the patient who decides to try to walk several miles at the beach 6 weeks after surgery. The fluid accumulation itself is not too concerning, but there is always the potential that the underlying cause may be infection, crystalline disease (gout or pseudogout – more on that in a moment), or other causes. Fluid aspirated from a knee replacement will usually be sent to the laboratory for analysis.

Pseudogout

Pseudogout is the accumulation of calcium pyrophosphate crystals within a joint. This can occur even without ever having knee surgery, and it looks similar to gout or an infection in the knee with redness, swelling, and pain. Surgery can sometimes trigger a pseudogout attack in the months after surgery. This is diagnosed by examining the joint fluid under a microscope and special light, identifying the characteristic crystals (gout can also affect the knee with a similar appearance, except the crystals are from uric acid in this case).

Compartment Syndrome

Compartment syndrome occurs when there is swelling within the muscles and soft tissues within an anatomic compartment that is so severe that it constricts blood flow, possibly enough to lead to death of the muscles and tissues. It is exceedingly rare to see after joint replacement surgeries and is in fact seen more often with traumas, particularly crush injuries to the lower leg in which there is extensive damage. It can occur rarely from other causes, such as a tourniquet that has been placed for too long around the leg during surgery with constricted blood flow. Another rare cause is from having saline that is normally pumped through the knee during arthroscopy leak out of the knee capsule into the tissues of the calf or thigh.

This is one of the reasons that we do not use tourniquets routinely for knee and ankle surgeries in our practice, but across the country there are many surgeons who do in order to decrease bleeding at the time of surgery. There is some debate as to whether the blood loss is really decreased, however, given that cut blood vessels will bleed after surgery, since the surgeon did not see them and cauterize them at the time of surgery. The high pressure tourniquet also can cause significant thigh pain after the surgery. However, it remains an accepted difference of opinion among surgeons and both techniques are accepted in the orthopaedic community.

Key Points For This Chapter:

- **The vast majority of patients undergoing orthopaedic surgery do very well (estimated 95%-98% without major complications)**

- **Surgery still has risks, and despite everything being done correctly to try to prevent them, complications can and do occur. Some are serious or life threatening.**

- **Complications include problems from anesthesia, medical complications, and surgical complications.**

- **In reality, there are more potential (and rare) complications that *can* occur than could be listed in this book or even several books, but the ones that occur most frequently are discussed here.**

- **Patients must be willing to accept the risks of surgery before deciding to have it performed.**

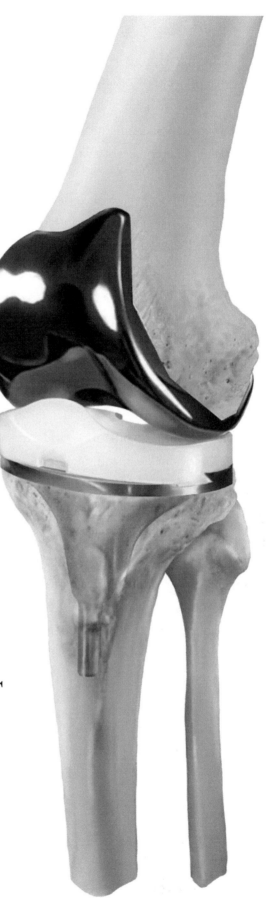

PART IV - REVISION HIP

AND KNEE REPLACEMENT

Chapter 37 - Revision Hip And Knee Surgery

Modern joint replacement surgery and the prostheses used in these surgeries are a mature technology, with many advancements over the past several decades. However, despite the wide success and excellent outcomes, nearly all artificial joints can be expected to wear out after enough time and wear. Joint replacement surgery is analogous to repairing the moving parts of a car, then sealing the engine compartment and achieving decades of 16+ hours of daily use without lifting the hood again!

Interestingly, many of the revision surgeries that we perform today are not so much because the implant failed, but rather because the bone surrounding the implant failed and no longer provides adequate support. The prostheses gradually may become loose.

There can be a number of reasons why joint replacements need to be revised. Trauma and periprosthetic fractures (fractures that occur around the replacement) often result from the same types of accidents that routine broken bones occur in, such as falling or being involved in a motor vehicle accident. Mechanical (aseptic) loosening is another common cause, as the prosthesis becomes loose from the surrounding bone; this often is an indirect result from a slow inflammatory reaction from worn away plastic material (osteolysis). Infection is another cause for revision surgery. Other potential causes include instability or recurrent dislocations, or mechanical failure, such as breakage of the artificial joint parts, although this has become a very uncommon reason in recent years when compared to those listed above.

Regardless of the reason, revision surgery is a complex and demanding surgery, with a wide variation in complexity depending on the diagnosis and underlying problem, scar tissue, potential damage to nerves or blood vessels (again, often because of scar tissue or calcified tissue called heterotopic bone), and longer surgical time and increased blood loss because of the need to remove the old components.

Revision joint replacements are among the most complex procedures in modern orthopaedics, and in contrast to routine primary (first time) joint replacement surgeries, relatively few surgeons and centers perform revision surgeries. For that reason, many of our patients undergoing revision surgery have been referred from other centers or surgeons.

Trauma and Periprosthetic Fractures

When a patient with an artificial hip or knee is involved in a fall or trauma, most often it is the surrounding bone that breaks before the metal prosthesis fails. As a result, the bone surrounding and anchoring the artificial joint may fracture, requiring that the joint replacement be revised as part of the surgery to treat the broken bones.

This often is a significantly more complex procedure than simply treating the fracture alone, as the old prosthesis may be loose but still attached in places to some bone fragments or cement. Old cement usually has to be completely removed from the bone, a time-consuming process that leads to extended operating time and increased blood loss. Scar tissue around the area leads to an increased incidence of nerve injury and bleeding.

Once the hip or knee replacement has been reconstructed with the surrounding fracture stabilized with fixation (often wires, plates, screws, and/or bone graft), it is common that a prolonged period of limited weightbearing is needed in order to allow the fractured bone surrounding the prosthesis to heal.

Mechanical (Aseptic) Loosening and Osteolysis

Another relatively common reason for failure of an artificial hip or knee is loosening of the bone surrounding the implant that occurs without any known infection. Over time, a patient's bone may simply re-absorb in the area that previously held the prosthesis with bony ingrowth or cement, leading to the gradual onset of pain as the prosthesis loosens.

A bone scan (not to be confused with a bone density scan, which checks for osteoporosis) is often helpful for diagnosing early loosening of a prosthesis that cannot yet be seen on regular x-rays. When loosening has been progressing for some time, advanced loosening changes become visible on regular

x-rays without the need for a bone scan.

A common reason for accelerated loosening is a process known as osteolysis. This term simply refers to the resorption of bone, but it usually is indirectly related to a slow inflammatory reaction caused by plastic wear. If an artificial joint contains a metal or ceramic surface that rubs against a plastic (polyethylene) surface, over a period of years millions of microscopic plastic particles are generated from the wear. White blood cells try to digest these plastic particles, and when they cannot be digested, the cells burst and release the enzymes that they normally use to digest bacteria and foreign bodies. As a result of this process being repeated millions of times, some of the bone that anchors the prosthesis in place is re-absorbed, leading to loosening.

As more has become known about this process, engineering efforts in the past decade have focused intensely on solving this problem. As a result, ceramic on ceramic or metal on metal joints have come into widespread use (especially for younger patients) that generate less wear by eliminating the plastic components.

In replacements that still need a plastic component, the plastic (polyethylene) has been improved markedly with materials engineering improvements, such as highly crosslinked polyethylene. These newer materials are thought to produce far less wear than previous generations of hip and knee replacements. Additionally, a sterilization process using gamma radiation in the 1990's was responsible for early wear of many of the plastic components within 10 years or less, but engineering advances in the past 15 years have largely made this a historical issue. The plastic components used today are expected to last many years based on simulator studies and retrieval studies.

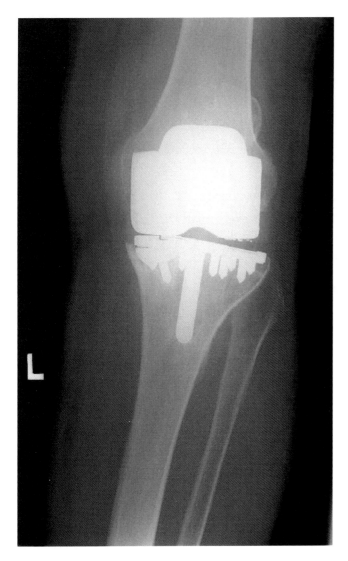

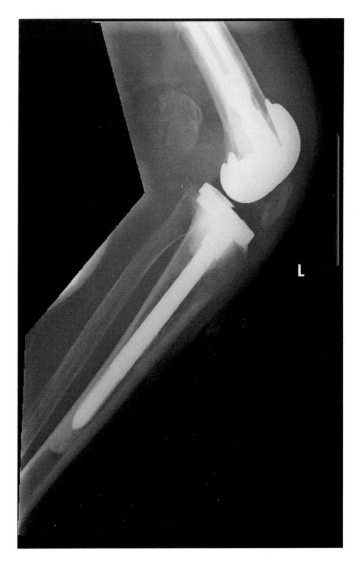

Figure 37-1. This patient presented with severe knee pain 20 years after his knee replacement surgery. The metal plate in the tibia is worn and cracked.

Figure 37-2. This is the same patient after revision of his knee replacement. This x-ray shows the knee from the side in order to better show the long stems used to reconstruct the knee after removing the old broken prosthesis.

Many modern hip and knee replacement designs allow for a limited revision surgery called a liner exchange. When the plastic liner in the hip socket or the plastic bushing between the metal parts of a knee replacement begins to wear out, often a limited surgery can be performed to simply replace the plastic component only.

In this way, the metal parts that are grown in or cemented to the surrounding bone do not need to be revised, and the limited revision has a fairly quick recovery and takes significantly less time than a surgery that revises the parts fixed to the bones.

Infections

Artificial joints do not often become infected, but when they do, surgery is usually required. Infections can either occur in the weeks or months after surgery or can occur because of hematogenous spread (spread of an infection through the bloodstream to the joint replacement). For this reason, patients with joint replacements are urged to seek medical treatment whenever they become ill with fevers or have a prolonged infection in another part of the body (such as a urinary tract infection or a diabetic ulcer on a leg that does not heal). There is a risk of a prolonged infection at one of these sites spreading through the bloodstream and causing the previously healthy joint replacement to become infected.

If an infection has only been present for a short period of time (a few days), it may be possible to simply "wash out" the joint. This is often referred to as "irrigation and debridement." This surgery usually involves cleaning the joint out, removing any infected appearing tissue, and washing a large amount of saline and antibiotics through the wound. Some infections may be treated with an arthroscopic washout, meaning that small incisions and an arthroscope may be used for the procedure.

When bacteria multiply and adhere to an artificial surface, it is difficult for the body's immune system to remove the bacteria. The most virulent types of bacteria multiply on the artificial joint surface and build a wall, called a glycocalyx, that prevents the immune system from getting at them. As a result, if an infection has been present for a while (more than a few days or couple of weeks), it may be necessary to remove the artificial joint entirely, treat the patient with antibiotics, and then re-implant the joint replacement weeks or months later when the infection appears resolved.

Often a cement spacer that is impregnated with high-power antibiotics is placed within a hip or knee joint after removing a prosthesis. This keeps the space for the replacement from filling in and also delivers antibiotics into the joint for weeks. The spacer is removed when the joint replacement is reimplanted.

It is not uncommon for an infection of any type (not just joint replacements) to require an open wound in order to heal. Infections that are closed over tend to not resolve as well as those that are left open, allowing the wound to drain. The body then fills in the infected space from the bottom up, with healing tissue called granulation tissue.

The classic treatment has employed frequent dressing changes to pack the wound as it heals, although recent advances have included wound vacs that have greatly improved the process. A surgical sponge is placed into the open wound and covered with a plastic dressing, and a vacuum pump is attached through a small hole in the dressing. This draws out any fluid but also pulls the edges of the wound close together, helping it to heal significantly faster. The vacuum pumps are small enough now that they may be worn on the patient's hip in a small holster, allowing the patient to get out and about while the vacuum pump heals the wound. The sponge and dressing typically only has to be changed two or three times per week in most cases, which has greatly facilitated the treatment of such wounds.

Instability or Dislocations

Sometimes a joint replacement becomes unstable, with the artificial joint dislocating or subsiding (sinking into the underlying bone that anchors it). Hip replacements can be particularly problematic if the ball keeps popping out of the socket (dislocating), which requires a trip to the hospital to pop the joint back into place under quick anesthesia. If it happens just once or twice, often as the result of some careless or risky activity, no further treatment may be necessary and the patient goes on to enjoy their joint replacement without difficulty.

At other times, however, the dislocations can become recurrent and problematic. Repeated dislocations (or problems with instability in general, as can occur with a knee replacement) may lead to the need to revise the joint replacement to make it more difficult to dislocate. For example, with a hip replacement that develops recurrent dislocations, possible treatments might include changing the ball to a larger diameter ball, using a liner in the socket that has a lip to add more stability (at a cost of decreased range of motion), changing the angle of the components if they are not optimal, or changing the length of the femoral neck to increase tension in the joint and hold it

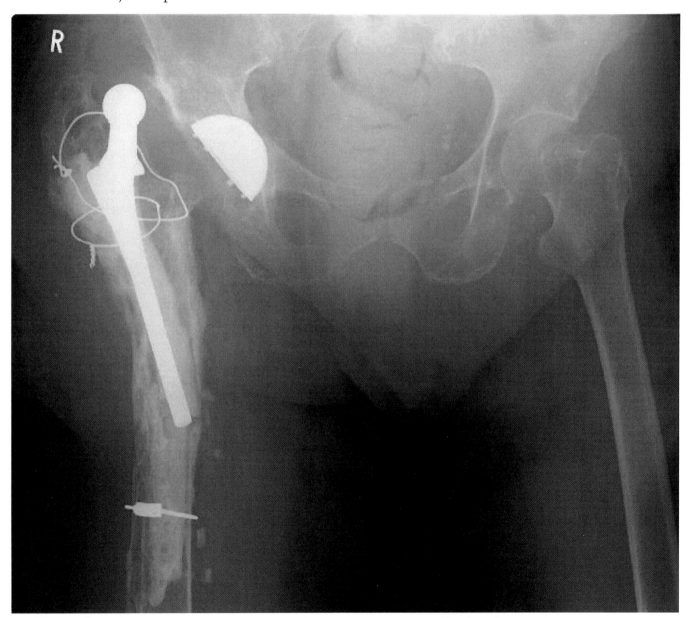

Figure 37-3. This patient presented to our practice having gone many years since she saw her surgeon (she had moved here from another city). At this point, this hip replacement is dislocated, chronically infected, and broken. This will require extensive surgery to fix, but most patients seek help years before reaching this point.

in place (with a trade-off of making the leg longer). Sometimes the underlying problem may be more complex, possibly including multiple causes, and extensive revision surgery may be recommended.

Outcomes From Revision Joint Replacement Surgery

As technology and surgical techniques have progressed, revision surgery has much better outcomes than it did twenty to thirty years ago. However, as noted at the beginning of this chapter, revision arthroplasty often is still one of the most complex procedures in modern orthopaedic surgery. Complication rates are higher among revision surgeries, particularly the more complex revisions that require extensive reconstruction or in older patients.

Not all revision surgery is necessarily complex; some revisions are limited and may only require exchanging a liner or other part. Still other surgeries may represent major undertakings to remove broken parts and cement, reconstruct portions of the pelvis, femur, or tibia, and augment with bone graft. There is a wide variation depending on exactly what is needed.

Key Points For This Chapter:

- Revision hip or knee replacement surgery can be quite complex

- There is a wide variation in the complexity of revision surgery, depending on the reason for the revision

- Although most joint replacements last for many years, there can be numerous reasons why revision may eventually be needed, including trauma or fracture, loosening of the prosthesis, infection, or instability (such as recurrent dislocations).

- Relatively few surgeons and centers offer revision joint replacement surgery

- Many revisions today are performed for joint replacements initially undergone one or two decades ago; with today's technology, we have good options for revisions of old replacements and also expect the first time joint replacement implants used today to last at least as long as those from the 1980's and 1990's (and likely longer)

APPENDIX I: RESOURCES

The best resource for information about orthopaedic surgery is (of course!) your surgeon. However, there are a number of additional resources available today that provide good information about orthopaedic surgery in general and total joint surgery in particular.

Here are a few of the better websites for more information:

The American Academy of Orthopaedic Surgeons

The AAOS is the principal organization for most practicing orthopaedic surgeons in America and is responsible for numerous publications and other channels for patient (and physician) education. The website is www.aaos.org.

The American Association of Hip & Knee Surgeons

The AAHKS is an organization devoted to hip and knee surgery. The AAHKS also provides an number of excellent articles and information sources. The website is www.aahks.org.

The Arthritis Foundation

This Arthritis Foundation is a large, nonprofit organization that provides support and information for all types of arthritis and arthritic conditions, with a plethora of information on nonsurgical treatments and news. The website is www.arthritis.org.

ActiveJoints.Com

A very informative website created by a someone who underwent hip replacement himself. The site has numerous links, lists of relevant books, and other resources. The website is www.activejoints.com.

APPENDIX II: DENTAL CARE

A frequently asked question is whether or not antibiotics need to be routinely taken for dental visits after hip or knee replacement.

It is possible, although very uncommon, to have bacteria introduced into the bloodstream during dental procedures that may infect an artificial joint. In previous years there have been conflicting recommendations by different orthopaedic surgeons and dentists. The American Dental Association and the American Academy of Orthopaedic Surgeons finally met together in 2002, had panel discussions, and composed a set of guidelines for patients, surgeons, and dentists. The following guidelines and recommendations are from the Academy and posted on their website.

When do you need preventive antibiotics?

You won't need to get preventive antibiotics for most dental procedures. But because you have an artificial joint, your risk of contracting a blood-borne infection is higher than normal. So preventive treatment is advised if the dental procedure involves high levels of bacteria.

You should get preventive antibiotics before dental procedures if:

- You have an inflammatory type of arthritis such as rheumatoid arthritis or systemic lupus erythematosis.
- Your immune system has been weakened by disease, drugs, or radiation.
- You have insulin-dependent (Type I) diabetes.
- You had a joint replacement less than two years ago.
- You have had previous infections in your artificial joint.
- You are undernourished or malnourished.
- You have hemophilia.

What procedures require preventive antibiotics?

You should get preventive antibiotics for the following procedures:

- Dental extractions
- Periodontal (gum disease) procedures
- Dental implant placement and reimplantation of teeth that were knocked out
- Endodontic (root canal) instrumentation or surgery
- Initial placement of orthodontic bands (not brackets)
- Injection of a local anesthetic into the gums near the jaw
- Regular cleaning of teeth or implants where bleeding is anticipated

What kinds of antibiotics are suggested?

The following preventive antibiotics are suggested:

- If you can take oral medications and are not allergic to penicillin, 2 grams of Amoxicillin, Cephalexin, or Cephradine should be taken one hour before the procedure.

- If you cannot take oral medications and are not allergic to penicillin, 2 grams of Ampicillin or 1 gram of Cefazolin should be administered by injection one hour before the procedure.

- If you are allergic to penicillin, 600 milligrams of Clindamycin should be taken orally or administered by injection one hour before the procedure.

These guidelines are designed to help doctors and dentists make decisions about preventive antibiotics for dental patients with artificial joints. It is not a standard of care or a substitute for the practitioner's clinical judgment, because it is impossible to make recommendations that would cover every situation. Practitioners must exercise their own clinical judgment in determining whether or not preventive antibiotics are appropriate.

APPENDIX III: CHECKLISTS

This section includes several sample checklists that may be helpful to patients as they prepare for office visits and hospitalization.

CHECKLIST – WHAT TO TAKE TO THE OFFICE VISIT

- Any prior x-rays, bone scans, or MRI's (both the actual films and reports)
- Any prior medical records that are pertinent (old operative reports, previous orthopaedic consultations)
- List of current medications and dosages
- Current insurance card / information
- Written list of questions to ask the surgeon
- Any family members or friends who will be involved in your care

CHECKLIST – WHAT TO TAKE TO THE HOSPITAL

- Any consent forms or other documents from your doctors
- Good nonskid slippers
- Pillow if you want
- Robe if you want
- Toiletries
- Reading material / portable music players / laptop computer
- Crutches or walker if you already have them

You probably should NOT take medications, credit cards, cash, jewelry, or pajamas to the hospital.

CHECKLIST – THINGS AT HOME AFTER SURGERY

- Crutches/walker
- Cane
- Extended grabber/reacher, sock tool, extended shoe horn
- Raised toilet seat if your toilets are low
- Shower stool
- Handheld showerhead on a cord
- Long-handled sponge
- Grab bars for showers are helpful
- Straight back chairs for sitting after total hip replacements
- Appointment cards

APPENDIX IV: ABOUT THE PRACTICE

Orthopaedics New England, is located in Middlebury, Connecticut, and focuses on hip and knee surgery as a specialty. The practice is internationally recognized for pioneering work in joint and adult reconstruction, and consists of two surgeons. Both trained at Yale University School of Medicine for orthopaedic surgery. Both Dr. John Keggi and Dr. Robert Kennon are board certified orthopaedic surgeons and actively involved in research, publications, and presentations at national conferences and symposia.

Dr. Kristaps Keggi was the founder of the practice and of the Keggi Orthopaedic Foundation, a nonprofit research and educational organization that has been responsible for numerous publications, research projects, and the education of hundreds of residents and visiting fellows from all around the world. He was the first the develop, publish, and use the modified anterior approach for hip replacement. He left the practice in 2009 when he was 75.

Dr. John Keggi is the nephew of Dr. Kristaps Keggi. He is fellowship trained in both adult reconstruction and pediatric orthopaedics, although his practice today focuses on hip and knee surgery. Dr. John Keggi and Dr. Robert Kennon together performed the first Birmingham hip resurfacing arthroplasty in Connecticut, and also jointly performed the first anterior Birmingham hip resurfacing via the minimally invasive anterior approach after developing the anterior procedure with cadaveric trials. He also serves on the board of JISRF.

Dr. Robert Edward "Ted" Kennon is also fellowship trained in adult reconstruction, focusing on hip and knee replacements, hip resurfacing, and knee arthroscopy. He also has a degree in mechanical engineering and is involved with orthopaedic research and publications that primarily focus on minimally invasive hip and knee replacement surgery, hip resurfacing, and related medical research. He is the author of this book.

APPENDIX V: CONTACT US

Our practice is located in Middlebury, Connecticut.

The primary hospital that we operate from is the Connecticut Joint Replacement Institute (CJRI) at Saint Francis Hospital in Hartford. Some surgeries are also performed at Waterbury Hospital in nearby Waterbury, Connecticut, and at the Naugatuck Valley Surgery Center also in Waterbury, Connecticut.

Many of our patients come from other cities, states, and countries for joint replacements and resurfacings. Because of the distance sometimes involved, we do offer telephone consultations regarding total hip replacement, hip resurfacing, partial and total knee surgeries, and revision surgeries. This does require that we have information from previous medical records, history forms that we mail or fax to patients, and x-rays and/or other imaging studies available for review prior to the telephone consultation. Inquiries should be directed to the assistants for either Dr. John Keggi or Dr. Robert Kennon.

Office visits are available both at the main office in Middlebury and at a satellite office in Hartford.

Patients wishing to make appointments may contact the appointment line via the main telephone number.

Orthopaedics New England
1579 Straits Turnpike
Middlebury, CT 06762

(203) 598-0700

Or visit us on the web at:

www.OrthoNewEngland.com

GLOSSARY

A

Abduction – refers to movement away from the midline of the body, such as lifting the leg out to the side. Opposite of adduction.

Acetabular cup – the artificial socket used in a hip replacement.

Acetabular liner – a liner that fits within the acetabular cup (socket) in a hip replacement. The liner may be plastic (polyethylene), metal (cobalt chrome), or ceramic.

Acetabulum – anatomical term referring to the socket of the hip joint.

Adduction – refers to movement towards the midline of the body, such as crossing the legs. Opposite of abduction.

Adult reconstruction – the subspecialty of orthopaedic surgery that deals with hip and knee replacements and related surgeries. Most specialists in adult reconstruction have fellowship training in this area after orthopaedic surgery residency.

Alumina oxide ceramic – the most common ceramic material used in artificial joint replacements. Ceramics have low friction and excellent wear resistance, but are expensive and in rare cases can fracture.

Anemia – having less red blood cells circulating than normal, resulting in fatigue and occasionally dizziness. Severe anemia can increase the risk of heart attack or stroke. Anemia can have many different causes, including blood loss from surgery, iron deficiency, B12 deficiency, and other causes.

Antalgic gait – refers to a type of limp that is present because of a painful joint (e.g., hip or knee).

Anterior – anatomical term referring to the front of the body.

Anterior cruciate ligament (ACL) – the ligament in the center of the knee that prevents the tibia from sliding forward under the femur.

Anterolateral – anatomical term referring to the front and side of the body, e.g., the anterolateral approach is from the side and slightly towards the front of the hip.

Antiemetic – a medication used to treat nausea and/or vomiting

Arthritis – inflammation of a joint (knee, hip, etc.)

Arthrodesis – fusion of a joint (causing the two sides to grow together into a single bone)

Arthrofibrosis – medical term for stiffness in a joint, often referring to scar tissue formation and tightness after surgery or injury

Arthrogram – any imaging study (such as x-ray or MRI) in which contrast is injected into a joint to yield a better picture of the joint and its contents

Arthroplasty – commonly refers to any surgery that replaces a joint

Arthroscope – an instrument used to look inside a joint. Modern arthroscopes use fiberoptic video cameras.

Arthroscopic drilling – surgical technique for drilling small holes in a cartilage "crater" in a joint, with the goal of causing the crater to fill in with smoothed scar tissue (similar to *microfracture*)

Arthroscopy – any surgery that involves looking inside a joint with a camera

Articular cartilage – the cartilage that coats the ends of joint surfaces and makes the joint move smoothly, analogous to Teflon coating in a frying pan. When it is gone, the resulting painful condition is literally "bone on bone" arthritis.

Aseptic loosening - loosening of an implanted joint replacement that occurs without infection

Aspiration/arthrocentesis – both terms refer to the procedure of drawing fluid out of a joint with a needle

Avascular necrosis (AVN) – also known as osteonecrosis. A pathologic disease process in which an area of bone begins to die, usually because of a lack of local blood supply. It can have many different causes and can occur in different parts of the body, commonly the femoral heads of the hips, the end of the femur in the knee, the talus in the foot, and the head of the humerus.

B

Bone graft – using bone or a bone-like material (such as coral) to fill in a defect or supplement an area of the body where its own bone is deficient. There are three general types of bone graft: autograft (a person's own bone, taken from somewhere else in the body), allograft (human bone from a cadaver), or xenograft (nonhuman bone).

Bursitis – an inflammation of a bursa, or sac-like structure between muscle layers. Bursae occur all over the body, commonly affected sites include the sides of the hips, the front of the knees, and the back of the elbow.

C

Cellulitis – an infection of the skin, usually superficial.

Chondromalacia – softening and scuffing of the articular cartilage that covers the bone in a joint. Can be a precursor to arthritis.

Chondroplasty – any surgical procedure that involves smoothing and/or reshaping the cartilage surface in a joint, usually referring to smoothing the roughened chondromalacia seen during arthroscopy

Cobalt chrome – a very strong alloy that has been used for many years in orthopaedics both for its strength and its biocompatibility in the body. Often used as a bearing surface.

Compression hip screw – a device used in the treatment of hip fractures to hold the fractured hip together until it heals. Screws can either be fixed with a compression buttress sideplate along the side of the femur or by a rod that goes through the center of the bone (a more modern and less invasive technique).

Computed tomography (CT) – Commonly referred to as a CT or CAT scan, this type of imaging produces many cross sectional images by passing the patient through a large spinning x-ray machine (looks like a giant doughnut).

Continuous passive motion (CPM) – an automated machine used by some surgeons to slowly move a joint (usually a knee after total knee replacement) back and forth, especially if a patient is unwilling or unable to move it on their own

Core decompression – a surgical procedure used to try to restart bone formation (especially in the hip) when avascular necrosis is causing it to die. The procedure involves drilling into the affect area to try to start a healing effect. It is not particularly effective and requires a prolonged period of limited weightbearing afterwards, but it is still often attempted in young patients with AVN to try to delay the need for joint replacement surgery.

Cortisone – a type of steroid injection. Cortisone is actually a brand name and is not commonly used today, but many patients and physicians still often refer to glucocorticoid steroid injections (such as into a joint or bursa) by this term.

Coumadin (warfarin) – a type of oral blood thinner. Originally developed over 50 years ago as a rat poison, many surgeons still use it to prevent blood clots after surgery. It requires careful monitoring with blood tests to check that the blood is not too thinned.

Crepitus – term referring to the crunching sound that a joint may make with motion (for a variety of different reasons, including arthritis)

Cruciate retaining knee – a type of knee replacement that keeps the posterior cruciate ligament (PCL) for stability, thought by many surgeons to lead to better preservation of joint biomechanics whenever it is possible to keep an intact PCL

Cryocuff – a sleeve device that fits around a joint (such as the knee) and can connect with a cooler to allow circulation of ice water, cooling the extremity/surgery site and helping significantly with postoperative discomfort and swelling

D

Deep venous thrombosis (DVT) – a potentially serious blood clot that forms within the deep returning veins of the leg and pelvis; the clot can break off and travel to the heart or lungs, resulting in death or other serious problems.

Degenerative joint disease (DJD) – a broad term describing a worn out joint, often due to osteoarthritis, posttraumatic arthritis, or other causes. Physical findings include stiffness, pain, and a roughened joint surface lacking sufficient smooth articular cartilage (seen on x-rays with decreased or thinning joint space, subchondral sclerosis, cysts, and osteophytes – spurs).

Dehiscence – term describing breakdown of a previously closed wound or incision.

Dislocation – what occurs when a joint is completely out of place (e.g., a shoulder or hip – or hip replacement – in which the ball slips out of the socket)

Doppler ultrasound (or DVT ultrasound) – a diagnostic imaging test that uses ultrasound to look inside the deep veins of the leg to see if there is a blood clot within the vessels (DVT)

E

ECASA – enteric coated aspirin, commonly used as a blood thinner for prevention of blood clots after surgery (thromboprophylaxis)

Edema – swelling in an extremity. There can be many causes for this, but some degree of edema is normal for the period following major joint replacement surgery.

Effusion – swelling in a joint. Arthritic effusions often result in "water on the knee," which is a painful, swollen joint that often benefits from aspiration and injection.

Enoxaprin – a low-molecular weight heparin (such as Lovenox) that is used as an injectable blood thinner to prevent blood clots (thromboembolism) after surgery

Epidural – a type of regional anesthesia that involves injection of anesthetic into the space around the spinal cord. It is safe and effective, offering several advantages over traditional general anesthesia. Epidural steroid injections are also sometimes used as a treatment for sciatica and other spine conditions.

Erythema – redness. Some erythema is normal after surgery and/or injury, but worsening and progressive erythema can be a sign of infection.

Extend – in the anatomical sense, straightening a knee or moving a hip backwards (opposite of flexion)

External rotation – in the anatomical sense, rotating the extremity outward (e.g., turning the toes outward)

F

Femoral condyle – the rounded ends of the femur (e.g., thigh bone) that make up the upper half of the knee. There are two condyles, one for each side of the knee (medial and lateral).

Femoral head – the "ball" of the upper end of the femur (e.g., thigh bone) that fits within the "socket" of the hip (acetabulum)

Femoral neck fracture – a type of hip fracture in which the narrow part of the thigh bone just below the "ball" (femoral head) is fractured. Minimally displaced neck fractures can sometimes be pinned percutaneously, and other femoral neck fractures require partial or total hip replacement.

Femoral nerve – the major nerve that runs along the inner thigh and makes the quadriceps and other knee extensors work

Femur – the "thigh bone," a long tubular bone with a ball at the top and the knee joint at the bottom

Ferrous sulfate – Iron pills, often used after surgery to help replenish the red blood cells. These do have a constipating effect, and laxatives and stool softeners are often prescribed with iron pills.

Fibromyalgia – a poorly understood pain syndrome that can lead to pain at "trigger points" and joints. There is no surgery for fibromyalgia, and most patients with this condition are managed by pain management specialists or rheumatologists.

Flex – in the anatomical sense, bending a knee backwards or bringing a hip upwards (as in to the horizontal position for the thigh)

Fluoroscopy – use of live x-rays for a procedure, such as injecting a hip joint or repairing a fracture

Foot drop – common term for a peroneal nerve palsy, a nerve injury that can result in inability to lift the foot up. If it persists or is permanent, an ankle foot orthosis (AFO) is used to keep from tripping on the downturned foot. It is a rare complication of total joint surgery that can result from injury or stretching of the sciatic nerve or its branches further down the leg.

Fracture – a broken bone. Note that a bone that is fractured IS broken (a common misunderstanding)

Fusion – See arthrodesis

G

General anesthesia – refers to anesthesia that affects the entire body, usually requiring a breathing tube, as opposed to a spinal or regional anesthesia

H

Hematocrit – a measure of the percentage of red blood cells occupying a volume of blood; used to determine degree of anemia and/or blood loss

Hemoglobin – similar to hematocrit, hemoglobin is a measure of the amount of functional red blood cells in the body. Patients with a low hemoglobin or hematocrit may need a blood transfusion.

Hematoma – a collection of blood, often resulting in an uncomfortable area of swelling. Hematomas can result from injury or surgery and may be made worse by excessive blood thinners.

Hip hemiarthroplasty – a partial hip replacement in which only the "ball" of the femur is replaced, without replacing the hip socket (acetabulum). This is typically used for very frail or ill patients with hip fractures.

Hip pinning – placement of percutaneous screws to treat a nondisplaced hip fracture

Hip resurfacing arthroplasty – an alternative to total hip replacement in which the femoral head is retained and a metal cap is placed over the end of the ball, rather than replacing the femoral side with a stem

Home exercise program (HEP) – a physical therapy regimen at home, typically used after surgery to maximize range of motion, strength, and function

Hyaluronate – also known as hyaluronic acid, a naturally occurring substance in the joints that is responsible for viscosity and lubrication. Hyaluronate injections are thick, clear gel injections used to cushion an arthritic joint.

Hyperemia – increased blood flow to an area, such as a knee after surgery, that results in redness and warmth

I

Ileus – a gastrointestinal complication after surgery in which the GI tract stops moving for a while, resulting in bloating, nausea, and vomiting. It usually resolves with stopping oral intake, placement of a nasogastric tube, and/or use of motility medications to start the GI tract moving again.

Incentive spirometry – use of a small breathing device after surgery to encourage deep breathing exercises and filling of the air sacs of the lungs. After surgery and anesthesia, closure of the small air sacs in the lungs (called atelectasis) is one of the most common causes of a fever, and deep breathing exercises help to avoid this.

Informed consent – concept of the patient giving permission for surgery after being informed of the common risks, benefits, and rationale for the procedure

Internal derangement – term used to describe a number of various conditions that result in mechanical problems with the knee, typically due to a torn meniscus, loose body, or disruption of the normal mechanics of the knee

Internal rotation – rotation of the leg/extremity inwards toward the center, often limited with hip arthritis

Intertrochanteric hip fracture – a type of hip fracture that occurs in the widest area of the upper femur, below the neck region. These types of fractures are typically treated with a dynamics hip compression screw or intramedullary hip screw rather than replacement or pinning.

Intramedullary hip screw (IMHS) – a device that is used to stabilize intertrochanteric hip fractures (also subtrochanteric and basicervical fractures), consisting of a rod that is placed through the center of the femur and transfixed with a lag screw into the femoral head

Iontophoresis – a physical therapy technique that uses an electrical current to introduce drug-carrying ions through the skin

L

Labrum – a gasket-like cartilage that surrounds the rim of the acetabular socket of the hip; tears can occur in the cartilage rim that result in pain and mechanical symptoms such as catching

Lateral collateral ligament (LCL) – one of the four main ligaments of the knee, the LCL is along the outer side of the knee and prevents buckling/instability from side to side

Lateral femoral cutaneous nerve – a sensory nerve that supplies the anterior and lateral (e.g., front and side) of the upper thigh region. It can sometimes be involved in temporary or permanent sensory changes after surgery, and rarely, meralgia parasthetica

Leg length discrepancy – having one leg longer or shorter than the other. It can be actual (e.g., the leg really is longer than the other) or apparent (e.g., leg lengths are equal, but pelvic tilt or obliquity makes them seem unequal)

Local anesthesia – anesthesia that relies on blocking the local nerves by injecting local anesthetics (similar to Novocaine)

Loose body – a free piece of tissue or foreign material (such as cement) that is floating about inside a joint, causing pain and mechanical symptoms (usually locking, catching, or buckling)

Lovenox – see enoxaprin

Lymphadema – swelling of an extremity (usually the legs or ankles) because of lymphatic congestion

M

Magnetic resonance imaging (MRI) – a method of imaging that uses magnetic fields instead of x-rays, producing detailed 3D images and cross-sectional images. It is particularly useful for examining soft tissues (muscles, nerves, cartilage)

Medial collateral ligament (MCL) - one of the four main ligaments of the knee, the MCL is along the inner side of the knee and prevents buckling/instability from side to side

Meniscal cartilage – see meniscus

Meniscectomy – procedure to remove a torn or damaged section of the meniscus to relieve knee pain and mechanical symptoms, usually performed arthroscopically

Meniscus - the C-shaped cartilage that fits like a gasket between the femur and tibia in the knee; it is often the damaged part responsible for internal derangement, commonly referred to as a "torn cartilage" of the knee. There are two of these inside each knee, both medial and lateral.

Meralgia paresthetica – a painful neuropathy that results from damage to the lateral femoral cutaneous nerve of the hip, typically producing burning pain over the front and lateral portion of the thigh

Microfracture – surgical technique for making small holes in a cartilage "crater" in a joint, with the goal of causing the crater to fill in with smoothed scar tissue (similar to *arthroscopic drilling*)

Modular total hip – a hip prosthesis that is assembled in sections, allowing greater flexibility in reconstructing the hip (e.g., better control of leg length, offset, anteversion or rotation of the hip)

N

Nerve palsy – term referring to an injured and improperly functioning nerve, can be either temporary or permanent

Neuropraxia – a temporary injury to a nerve, usually caused by stretching, which usually resolves over time

NPO – Latin term ("Nil per os") meaning "nothing per mouth," commonly ordered in the hospital for patients undergoing surgery in the morning who should not have anything to eat or drink in the hours before surgery

NSAIDS – Nonsteroidal antiinflammatory agents, or medications that reduce inflammation such as ibuprofen, naprosyn, or aspirin

O

ORIF (open reduction and internal fixation) – any surgery in which the fracture is opened, put back in place (reduced), and stabilized with orthopaedic hardware such as a plate and screws

Osteoarthritis – degenerative arthritis of the joints caused by breakdown and loss of the cartilage covering the ends of the joints, commonly referred to as "wear and tear" arthritis. It is the most common form of arthritis.

Osteochondral defect – a "pothole" in the surface of a joint, either due to trauma or biologic causes

Osteochondritis dessicans – disease in which the articular cartilage covering the surface of the bone begins to thin unexpectedly

Osteolysis – resorption of bone, commonly the bone surrounding an artificial joint implant and leading to early loosening of the prosthesis

Osteonecrosis – see avascular necrosis (AVN)

Osteophyte – a bony spur around the edge of a joint, commonly seen in arthritic conditions

Osteoporosis – loss of bone density. There is normal bone tissue, just not enough of it, leading to increased risk of fracture.

Osteotomy – any surgical procedure that involves cutting the bone (and usually resetting it to change alignment)

Oxinium – a proprietary material developed by Smith & Nephew in which a zirconium oxide (ceramic) coating is formed over a metal bearing surface by high heat and pressure.

P

PACU (post-anesthesia care unit) – another term for the recovery room, where patients are closely observed for a while after surgery until stable

Patella – the "knee cap," a thick rounded bone that essentially works as a pulley that allows the extensor mechanism to stretch over the front of the knee.

Patellofemoral arthroplasty – a partial knee replacement that resurfaces only the undersurface of the patella (knee cap) and the groove in the femur in which it travels.

Patellofemoral joint – the space between the patella (knee cap) and the femur

PATS (pre-operative admissions testing) – a series of tests performed prior to surgery, typically including an EKG, blood tests (coagulation profile, basic electrolytes and chemistries, complete blood count to check hemoglobin levels, etc.), blood donation and/or typing to match for any needed banked blood, chest x-rays, and other basic tests. The exact tests vary from patient to patient (depending on age and other medical diagnoses) and from institution to institution.

Peripheral vascular disease – a family of circulatory diseases. Poor circulation and blocked arteries in the legs can lead to an increased incidence of complications and problems, and sometimes this can be severe enough to limit options for surgery. It can also be directly responsible for pain in the legs and calves (called claudication).

Periprosthetic fracture – a fracture of the bone that occurs close to or around an artificial joint. These are typically seen as the result of a fall or trauma and often require surgery for treatment.

Phonophoresis - a physical therapy technique that uses ultrasound to introduce therapeutic drugs through the skin

Physical therapist (PT) – an allied health professional who specializes in exercises, stretches, and other modalities (such as ultrasound or electrical stimulation) to treat injuries and patients who have had therapy ordered by their surgeon after surgery.

Physician assistant (PA) – a health professional who assists physicians by performing physical examinations, rounding in the hospital, and performing minor procedures such as injections or setting fractures. Physician assistants perform their work under the supervision of a physician.

Polyethylene liner – the high performance plastic liner that is often used between metal parts in a hip or knee replacement.

Polyethylene wear – wear of the plastic liner in a hip or knee replacement (not all prosthesis designs include this type of bearing), which usually occurs over a period of many years.

Polymethylmethacrylate cement – bone cement used to hold orthopaedic hardware, such as knee replacements, in place and attached to the bone.

Porous coated – as an alternative to cement, some prostheses have a porous coating that the patient's bone grows into. This provides a stronger interface without the concerns for cement loosening, but requires stronger bones and is not at maximum strength immediately after surgery.

Posterior – anatomic term referring to the back side of the body (for example, the buttock area when discussing posterior approaches to the hip).

Posterior cruciate ligament (PCL) – the ligament in the center of the knee that prevents the tibia from sliding backward under the femur.

Posterior stabilized knee – a type of total knee replacement that replaces the function of the posterior cruciate ligament (PCL) by using a plastic post in the center of the artificial knee.

Posttraumatic arthritis – a form of degenerative joint disease that occurs because of an old injury or trauma. In its advanced stages, it is very similar to osteoarthritis.

Preoperative clearance – the process of checking that a patient is healthy enough to undergo surgery and anesthesia, usually consisting of a history and physical exam by the patient's primary care provider or a hospital internist. Sometimes underlying medical conditions such as diabetes or high blood pressure are discovered (or found to not be optimized), and these are corrected prior to surgery.

Prophylactic antibiotics – antibiotics given just before or during surgery to prevent infection.

Pulmonary embolus (PE) – a blood clot (thromboembolus) that travels to the lungs

R

Radiculopathy – disease or compression of the spinal nerve roots, often because of a bulging disc, that causes pain, numbness, and/or loss of motor function in an extremity. The most common scenario is a bulging disc in the low back that causes numbness, tingling, and pain down the back of the legs down to the foot (sciatica).

Radiographs – another term for x-rays.

Range of motion (ROM) – the limits to which a joint may move (e.g., flexion and extension limits of a knee).

Regional block – a type of anesthesia that provides for local, reversible loss of sensation and/or motor function, often used when general (whole body) anesthesia is unnecessary or undesirable

Revision – refers to a "redo" surgery, such as a revision joint replacement for a worn out joint replacement. These are typically more complex than first time (primary) surgeries and are usually only performed at specialized centers.

Rheumatoid arthritis – a type of inflammatory arthritis that is caused by the body's own immune system attacking the joints.

S

Sciatica – see radiculopathy.

Sepsis – term referring to infection. Generalized sepsis refers to infection in the bloodstream and is quite serious.

Septic arthritis – infection of a joint, such as a hip or knee. This usually requires surgical treatment, often as an emergency if the patient is quite ill (septic) from the infection.

Short-term rehabilitation (STR) – a short stay of several days to a few weeks at a rehabilitation center for physical therapy, until a patient is ready to function independently and safely in their own home.

Sinus tract – an opening that forms to the skin from an area of deep, chronic infection.

Spinal anesthesia – a type of regional anesthesia in which an injection of medication into the space around the spinal cord provides temporary and reversible sensory loss.

Staples – a type of wound closure that employs surgical staples to close the skin. These are typically removed at 7 to 14 days after surgery in most cases.

Steristrips – small adhesive bandaid-like strips used to strengthen a closed incision, usually used in conjunction with an absorbable suture.

Subtrochanteric hip fracture – a type of hip fracture that occurs just below the flared area of the upper femur, adjacent to the shaft. These typically are treated with an intramedullary hip screw in most cases.

Sundowning – a term used to refer to temporary confusion that is often worse at night, typically seen in elderly patients or patients with early dementia, who become easily disoriented with pain medications and anesthesia.

Supracondylar femur fracture – a shaft fracture that occurs just above the knee.

Synovial fluid – the body fluid within a joint. A small amount is normal; large amounts can cause painful swelling and may need to be drained.

Synovium – the tissue that forms the lining of a joint, normally responsible for making synovial fluid. It can become inflamed in arthritic conditions or injuries.

T

Thromboembolic disease – the formation of blood clots within the veins. This can lead to problems such as deep vein thrombosis (DVT) or pulmonary embolus (when a blood clot travels to the lungs).

Thromboprophylaxis – treatment to prevent the formation of blood clots after surgery or injury. Several methods are available, including mechanical prophylaxis (using pneumatic pumps on the feet or calves to promote blood flow) and pharmacological prophylaxis (blood thinners). One of the best methods of prevention is early mobilization and walking.

Tibial plateau – the joint surface of the knee that is formed by the upper end of the tibia.

Time-out – a step in the operating room just before the start of the procedure, in which the surgeon, anesthesiologist, and nursing staff confirm the identity of the patient, procedure being performed, and the the surgical site (usually marked by the patient and/or surgeon before surgery in the preoperative interview).

Titanium – a very strong and light metal alloy used for many types of orthopaedic implants.

Total hip arthroplasty – an artificial joint replacement in which both the "ball" (femoral head) and "socket" (acetabulum) are replaced.

Total knee arthroplasty – an artificial joint replacement in which the ends of the femur and tibia are replaced. The undersurface of the patella (knee cap) may or may not be replaced.

Transfusion – giving blood products to a patient through an I.V. Note that red blood cells, plasma, platelets, and other blood products (cryoprecipitates) can be transfused individually or together (whole blood). Autologous transfusion is giving the patient back his or her own blood, and allogeneic transfusion is transfusion of blood from a donor.

Trochanteric bursitis – inflammation of the bursa, or sack-like structure between muscle layers over the side of the hip. It often leads to pain over the side of the hip that can be reproduced by pressing or lying on that side.

U

Unicondylar knee arthroplasty – a type of partial knee replacement that only replaces one side of a knee (usually the medial, or inner, side of the knee).

Urinary retention – condition of being unable to void or urinate. Many factors can cause this, such as narcotics or spinal anesthesia, enlarged prostate, or other conditions. It may require temporary placement of a catheter.

V

Valgus – anatomic term for outward angulation of a limb away from the midline (e.g., knock-kneed). Opposite of varus.

Varus – anatomic term for inward angulation of a limb towards the midline (e.g., bowlegged). Opposite of valgus.

Visiting nurse agency (VNA) – the nurse and staff that visit a patient at home when he or she is unable to leave the home.

INDEX

A

Alumina oxide ceramic, 28, 32,33,45-49, 112,169,181

Anemia

preoperatively,137,140

postoperatively, 158

Anterior approach

for hip replacement, 36-39,59,179,181

for hip resurfacing, 34, 36-39

for knee surgery, 103

Anterior cruciate ligament (ACL) –

anatomy of, 71,93,112

injury to, 81,85

Anterolateral approach to the hip, 36

Antiemetic, 158

Arthritis, pathology of

Hip

Osteoarthritis, 10-11

Rheumatoid, 11-12

Posttraumatic, 13-14

Knee

Osteoarthritis, 75-77

Rheumatoid, 78

Posttraumatic, 77-78

Arthrodesis (also known as fusion)

of Hip, 40

of Knee, 107

Arthrofibrosis of knee, 183

Arthrogram

of Hip, 18

of Knee, 85

Arthroplasty

of Hip,

Indications for, 26, 133-135

Technique for, 27-29

Implants used for, 44-50

Complications of, 181-183

Total, 27-29

Partial (hemiarthroplasty), 29-30

Resurfacing, 31-34

of Knee,

Indications for, 97, 133-135

Technique for, 98-100

Implants used for, 110-114

Complications of, 183-186

Total, 98-100

Partial (unicondylar), 100-102

Patellofemoral, 102

Arthroscopic drilling, 81,93,96

Arthroscopy

of Hip, 42-43

of Knee, 79-80, 92-96

Articular cartilage

of Hip, 6

of Knee, 70-71,75

Aseptic loosening of implants, 45,47,65,86,100, 124,180

Aspiration/arthrocentesis

of Hip, 20

of Knee, 86

Avascular necrosis (AVN)

of Hip, 7,12,17-18,41-42,66

of Knee, 80-81,85

B

Birmingham Hip Resurfacing, see hip resurfacing arthroplasty

Bone graft, revision arthroplasty and, 27-28, 113,191-195

Bursitis

Trochanteric (hip), 7,10,15-16,23,179-182

Pes anserine (knee), 75,81

Prepatellar (knee), 75, 81

C

Cellulitis, 177

Chondromalacia, 71,79

Chondroplasty, 79

Cobalt chrome, 31,45-48,111-112,125

Compression hip screw for hip fracture, 13-14

Computed tomography (CT), 19,85,108,164

Continuous passive motion (CPM), 116,121

Core decompression, for hip AVN, 41-42

Cortisone – see steroid injections.

Coumadin – see warfarin

Crepitus, 180

Cruciate retaining knee replacement, 110-112

D

Deep venous thrombosis (DVT)

prevention of, 162,164-

complication of, 8, 163-164,174

Degenerative joint disease (DJD) – see arthritis

Dehiscence of a wound, 177

Dental vists, prophylaxis and, see appendix II

Dislocation of a hip, 28,46,55,65,180

Doppler ultrasound (or DVT ultrasound), 20,86, 163-164

E

ECASA, enteric coated aspirin

stopping before surgery, 147

for DVT prophylaxis after surgery,8,95,165

Edema, 18,85,1641,76

Effusion of the knee, 77,88,185

Enoxaprin (lovenox), for DVT prophylaxis after surgery, 8,52,116

Epidural anesthesia, 155

Erythema – redness, as a sign of infection, 131

F

Femoral condyle, anatomy of, 81

Femoral head, anatomy of, 6-7,12,26

Femoral neck fracture of the hip, 13,33

Femoral nerve

anatomy of, 8

complications with, 179

Ferrous sulfate (Iron pills), after surgery, 140

Fluoroscopy, for hip injection, 23

Foot drop see nerve palsy.

Fracture

   of hip as trauma, 13-14

   of hip/knee as complication, 42,66,178

   periprosthetic, 125,190-91

Fusion – See arthrodesis

G

General anesthesia, 51,115,151,155-157

H

Hematocrit, testing of, 158

Hemoglobin, testing of, 158

Hematoma, complication of, 165-166,178

Hip hemiarthroplasty (a partial hip replacement), for hip fractures, 29-30

Hip pinning, for hip fractures,13-14

Hip resurfacing arthroplasty,

   Indications for, 31

   Advantages of, 32-33

   Disadvantages of, 33-34

   Technique of, 31-32

Home exercise program (HEP),

   for hip replacement/resurfacing, 55-62

   for knee replacement, 119-122

Hyaluronate as injections for knee,89

I

Ileus as complication,175

Informed consent before surgery, 138-139

Internal derangement of knee, 80,92,185

Intertrochanteric hip fracture,13-14

Intramedullary hip screw (IMHS),13-14

L

Labrum, hip, tears of, 6,42-43

Lateral collateral ligament (LCL) of knee,72, 81, 112

Lateral femoral cutaneous nerve, 8,38,179

Leg length discrepancy, 182

Local anesthesia, 93-94,156

Loose body of the knee, 78,92,108,185

Lovenox – see enoxaprin

M

Magnetic resonance imaging (MRI)

   for hip problems, 18-19

   for knee problems, 80,84-85

Medial collateral ligament (MCL) of knee,72, 81, 112

Meniscal cartilage – see meniscus

Meniscectomy, partial arthroscopic, 94

Meniscus of the knee,

   anatomy of, 72

   tears of, 75,79-80,92-94

   arthroscopy for, 93-94

Meralgia paresthetica, complication of,8,38,179

Microfracture, arthroscopy and, 81,93,96

Modular total hip, prosthesis design, 29,44,48

N

Nerve palsy, 8,178

Neuropraxia, 95,179

NPO, before surgery, 148

NSAIDS, Nonsteroidal antiinflammatory agents, used for arthritis pain and, 24,90,145,176

O

ORIF (open reduction and internal fixation), 11-15

Osteoarthritis – see arthritis.

Osteochondral defect, of knee, 75,78-79,93

Osteolysis, complication of, 47-48,180,190

Osteonecrosis – see avascular necrosis (AVN)

Osteophyte seen with arthritis, 7,17,26,77,84,97

Osteoporosis, 13-14,29,32,45,89,134

Osteotomy

of the hip, 36,41

of the knee (high tibial), 107-108

Oxinium, bearing surface,

for hip replacements,48-49

for knee replacements, 99,112

P

PACU (post-anesthesia care unit, or recovery room), 51,95,115,144,152

Patellofemoral arthroplasty, 102,111

PATS (preoperative admissions testing), 136-

137,140-141

Peripheral vascular disease of the leg, 83

Periprosthetic fracture, see Fracture.

Physical therapy (PT), 15,23,51-52,59-62,71,73,81, 88-89,115-117,118-119,145,158-161,179

Physician assistant (PA), 130,138,144

Polyethylene liner

of hip replacements,28,33,46

of knee replacements,98,111,125

Polyethylene wear of components,33,46-47,125, 180,191-192

Polymethylmethacrylate cement (bone cement),28,44-45,100,111

Porous coated components, 27-28,44-45,52,100, 111, 180,182

Posterior cruciate ligament (PCL), 71,78,81-82, 112

Posterior stabilized knee, prosthesis type, 112

Posttraumatic arthritis – see arthritis.

Preoperative clearance before surgery,136-137, 143,147-148,153,173

Prophylactic antibiotics,

before dental procedures, 66,124,197-198

before surgery, 95,177

Pulmonary embolus (PE), 8,162,164

R

Radiculopathy, lumbar, as source of hip pain,10, 23,179

Radiographs (x-rays), 17-18,30,83-84,131

Regional block, for anesthesia, 144,148,156-7,165

Revision arthroplasty,

reasons for, 29,33,41,45,48,65,124-125,180, 184, 190-194

risks of, 28,45,74,105,140,178-179

outcomes of,48,102,113,194-195

Rheumatoid arthritis – see arthritis.

autologous, 141

donated, 139,140-141

Trochanteric bursitis of the hip, 7,15,182

S

Sciatica – see radiculopathy.

Sepsis, complication of (infection), 12,80,177

Septic arthritis – see sepsis.

Short-term rehabilitation (STR), after surgery, 52, 54, 117,138,160-161

Spinal anesthesia,

advantages of, 151,,153-154

disadvantages of, 154-155

Steroid injections,

for trochanteric bursitis, 15,23

for hip arthritis, 21,23-24

for knee arthritis, 81,88-89

Subtrochanteric hip fracture, 13-14

Sundowning (mental status changes), after surgery, 155, 175

Supracondylar femur fracture, see Fracture.

Synovial fluid and lining, 78,89,92

U

Unicondylar knee arthroplasty – see arthroplasty.

Urinary retention, as surgical complication,

V

Visiting nurse agency (VNA), after surgery, 52, 117, 138, 145

X

X-rays, see radiographs.

T

Thromboembolic disease

definition of, 8,95,162,174

preventing, 154,160,163-167

treating, 167

Tibial plateau, fracture of, see Fracture.

Titanium – in implants, 45,50,111,182

Total hip arthroplasty – see arthroplasty

Total knee arthroplasty – see arthroplasty

Transfusion